ECONOMIC COMMISSION FOR EUROPE
Geneva

Structural changes in international steel trade

Prepared by the Steel Section of the ECE Industry and Technology Division

UNITED NATIONS
New York, 1987

NOTE

The designations employed and the presentation of the material in this publication do not imply the expression of any opinion whatsoever on the part of the Secretariat of the United Nations concerning the legal status of any country, territory, city or area, or of its authorities, or concerning the delimitation of its frontiers or boundaries.

ECE/STEEL/54

UNITED NATIONS PUBLICATION

Sales No. E.87.II.E.33

ISBN 92-1-116408-7

02300P

PREFATORY NOTE

The Steel Committee, at its forty-ninth session, in October 1981, decided to include in its programme of work the study of structural changes in international steel trade (ECE/STEEL/34, para. 37). Active work on the project was started in February 1984.

The topic of the study was selected in the light of the structural changes which had taken place since 1970 in international trade in steel. The study, based on the statistical data presented in the UN/ECE publications statistics of World Trade in Steel, complemented by the United Nations COMTRADE (Commodity Trade Statistics) data base, covers the period from 1970 to 1982, although some tables also give statistical data on international steel trade for 1983 and 1984.

Chapter I describes the methodology applied in the preparation of the study.

Chapter II analyses the structural changes which occurred during the period in exports of steel products from various countries and/or regions.

Chapter III analyses the structural changes in imports of steel products by various countries and/or regions.

Chapter IV analyses the changes in foreign steel balances of countries and/or regions.

Chapter V examines the changes in the coefficients of specialization of exporting countries and/or regions in various importing zones.

CONTENTS

CONTENTS (continued)

LIST OF TABLES

LIST OF TABLES (<u>continued</u>)

LIST OF TABLES (continued)

LIST OF TABLES (<u>continued</u>)

LIST OF TABLES (<u>continued</u>)

LIST OF TABLES (<u>continued</u>)

LIST OF TABLES (continued)

LIST OF FIGURES

List of Figures

INTRODUCTION

The aim of the present study is the analysis of the structural changes which took place in international steel trade between 1970 and 1982. Very important economic changes occurred during this period. Amongst the most important, mention should be made of the dramatic increase in oil prices in 1973 and 1979 and the long-term impact of those increases on world economic growth and on steel consumption.

These economic changes made demands on iron and steel industries which differed considerably from those of the previous decades. At the same time, possibilities arose for reconstruction and modernization which could satisfy those changed requirements.

Restructuring in the industry embraces a wide range of activities throughout the world. The adjustment process differs from country to country, both in nature and in degree. It has been more successful in some countries than in others. In western Europe, it has taken the form of modernization and the closing down of facilities to bring capacity into line with the reduced demand for certain products. This is generally referred to as rationalization of production. It should be mentioned that, in western Europe, efforts have also been directed at improving quality in steel products and increasing labour productivity. In the United States the objective has been somewhat different: while some plants have closed down, the principal aim has been to modernize and expand capacity with an eye to the future. Japan has been engaged in a modernization programme aimed not at expanding productive capacity but at improving the quality of products and productive efficiency. The slow-down in the growth of iron and steel production in the USSR and other countries of eastern Europe reflects the policy followed, whereby rapid expansion has been dropped in favour of better and more productive use of existing installations.

World economic development in general has affected the global demand for iron and steel, and one of the long-standing features of the steel industry has been an increasing international trade in semi-finished and finished steel products. As a result, many national steel industries have carried a varying capacity of "export production", a factor which has figured prominently in many of the planning decisions regarding national steel capacity. The real significance lies not in the growth of the volume of trade itself, but in the intensity in the steel trade of individual producers which has evolved over the period from 1970. For instance, Japan was exporting more than 30 per cent of its steel production whilst importing around 1 per cent during the major part of the 1970s. On the other hand, the United States had become the world's largest steel importer, importing the equivalent of approximately 20 per cent of domestic production and exporting some 2 to 3 per cent.

According to a concept which has been used for a number of years, the functions of international steel trade can be divided into two main types:

(a) Steel trade to cover a deficit in steel supply where this deficit arises from the lack of domestic production facilities relative to domestic demand at the time of trade. Deficits in supply are expressed by the excess of steel consumption over production, which is equal to net imports. At the world level, the sum of net imports is equal to the sum of net exports, and they represent the total shipments for covering the deficit; and

(b) Comparative advantages in the production of particular grades or types of steel lead to exchange of steel products between countries. Such an exchange of steel products may occur even where these countries have the capability to meet all or nearly all of their steel requirements for domestic use. The difference between total deficit coverage and total world trade is the volume of exchange trade.

Although this is perhaps an over-simplification, these two factors do represent a broad distinction which is useful when considering aggregate levels and product patterns in international steel trade. It should also be remembered that the two factors are not mutually exclusive; both may contribute simultaneously to steel trade between individual nations.

Exchange trade is increasingly becoming the driving force for the considerable and continuing expansion of world trade in steel. Moreover, as developing nations expand their steel production capacity, it seems likely that exchange trade will continue to increase in importance. Indeed, many developing nations see the steel industry as an important element in their overall national economic strategy.

The structure of international steel trade during the period also reflected the existence of regional economic groups (free trade areas, the European Economic Community (EEC) and the Council for Mutual Economic Assistance (CMEA) and other forms of regional economic co-operation). Thus, intra-regional trade within EEC and CMEA accounted alone for nearly 30 per cent of world trade in steel (table 0.1).

The period since 1970 has seen a vigorous growth of steel trade. At the same time, there appears to have been an increasing degree of international specialization, in particular in steel products and steel grades. This would seem to be the case for steel tubes, large-diameter pipeline steel, special construction steels and certain grades of alloy steel.

Steel trade consists, of course, in a two-way flow - exports and imports. It is important therefore to consider the net results of steel trade when looking at the overall position of any national steel industry. In the main industrialized countries, national levels of crude-steel production and of steel consumption measured in crude-steel terms vary along parallel trends. In western Europe, the United States and Japan, however, the level of domestic crude-steel production differs significantly from the corresponding level of domestic steel consumption, thus affecting the importance of world trade in steel.

Since the purpose of this study is to analyse the structural changes which have occurred in international trade in semi-finished and finished steel products, indirect trade in steel is not taken into consideration. However, it should be mentioned that the pattern and volume of world trade in manufactured goods containing steel are of considerable importance to the steel industry. In many countries, the domestic consumption of steel includes a substantial proportion of steel which, at a later stage, will leave the country in the form of manufactured goods containing steel.

It would seem that the distinction between deficit coverage and exchange relations which was made above for direct steel trade is equally relevant to indirect trade.

World steel production, consumption and international steel trade

Steel production can be measured at a number of points along the production process. The first measurement may be taken immediately after tapping from a steelmaking furnace. This is generally referred to as "liquid-steel production". The second measurement point follows the solidification stage of the liquid steel, the total tonnage measured at this stage being collectively defined as "crude-steel production". Crude-steel production can be recalculated into "steel production in ingot equivalent". The final commonly used measurement point occurs at the conclusion of all steel plant operations, where steel production is measured in "finished steel product tonnes". Such a figure represents the weight of finished steel products in the form in which they leave the steelworks.

Throughout this study, all steel tonnages refer to production in ingot equivalent and take into account the improvements in yield generated by the introduction of continuous casting, calculated as follows: steel production in ingot equivalent = crude-steel production + 0.175 x continuously cast output.

Almost all international steel trade statistics are expressed in tonnes of finished products. In order to compare international steel exports and imports with steel production in ingot equivalent, trade statistics have to be multiplied by a factor taking into account losses incurred in producing the particular products. These factors vary according to product categories. Ideally, therefore, for each product, a specific conversion factor should be used to obtain the ingot equivalent of the product. Since, in the case of the present study, data difficulties precluded such an approach, an average conversion factor had to be used. The trade in ingot equivalent was calculated as follows: trade in ingot equivalent = trade in product tonnes multiplied by a uniform conversion factor of 1.30.

Table 0.1 shows world steel production and total world exports of steel products, as well as world exports of steel products in percentage of world steel production.

The volume of world steel production grew during almost the entire period under review. The trend, however, was not linear, owing primarily to the influence of the world economic recession and to cyclical fluctuations. The greater part of this expansion occurred from 1970 onwards, production growing from 600.23 million tonnes in 1970 to a peak level of 777.46 million tonnes in 1979. While, in terms of absolute tonnages, the greatest increase in production occurred in industrialized countries, from 1970 to 1982 the greatest relative growth took place in the new steel-producing countries of Latin America, the Far East and Asia.

Although it is only to be expected that the highest growth should be in new steel-producing countries, their initial level of production and the actual tonnages they produce must be taken into account.

World steel exports grew from 116.93 million tonnes in 1970 to 167.98 million tonnes in 1974; following a decrease in 1975 (145.80 million tonnes), they grew again up to 1979, when they reached their peak (181.23 million tonnes), and declined to 169.74 million tonnes in 1982. However, it can be stated that, in spite of fluctuations, the volume of world trade in steel products had a growing trend.

Table 0.1. World steel production and export

(Millions of tonnes in ingot equivalent)

Year	Production	Exports	Exports in percentage of production
1970	600.23	116.93	19.5
1971	587.63	124.41	21.2
1972	638.03	132.27	20.7
1973	709.44	146.27	20.6
1974	722.87	167.98	23.2
1975	661.01	145.80	22.1
1976	694.95	162.27	23.3
1977a/	696.64	163.71	23.5
1978	742.32	176.76	23.8
1979	777.46	181.23	23.3
1980	752.34	178.09	23.7
1981	746.47	181.13	24.3
1982	685.99	169.74	24.7

a/ USSR exports for the period 1977-1982 have been estimated, as described in chapter I.

World steel exports as a percentage of world steel production had grown, with small fluctuations, from 19.5 per cent to 24.7 per cent during the period 1970-1982.

Table 0.2 shows the ratio of total steel production to total steel consumption for selected countries or regions over the period 1970-1982. The value of the ratio equal to 1.00 indicates overall equality between production and consumption. As this value increases above 1.00, it indicates an increasing reliance on a surplus of exports over imports in maintaining current steel production levels. If the value declines below 1.00, it signifies an increasing dependence on imports in meeting domestic steel demand. This table shows that, in 1970, the main feature was the surplus of exports over imports in Japan, EEC and the USSR and the enormous deficit in such regions as Africa and the Middle East. By 1982, the situation had not changed significantly and was dominated by Japan's surplus. EEC had an even larger surplus, exceeded only by Japan, while Africa (excluding South Africa) and the Middle East still had the largest deficit. The other western industrialized countries had moved from a deficit to a marginal surplus but it should be mentioned that this improvement may be attributed particularly to such countries as Austria, Spain, Sweden. As for the Far East and Latin America, the situation had not improved considerably in spite of the fact that such countries as the Republic of Korea and Brazil had become net exporters.

Table O.2.   Ratio of total steel production to total steel consumption

	1970	1971	1972	1973	1974	1975	1976	1977	1978	1979	1980	1981	1982
Africa a/	0.13	0.16	0.18	0.18	0.14	0.18	0.21	0.16	0.16	0.15	0.16	0.16	0.16
Far East	0.74	0.72	0.74	0.70	0.68	0.74	0.72	0.72	0.70	0.71	0.78	0.80	0.83
Japan	1.32	1.50	1.37	1.35	1.51	1.53	1.69	1.65	1.56	1.45	1.43	1.44	1.44
Oceania	0.96	0.86	1.00	0.99	0.92	1.13	1.18	1.28	1.34	1.25	1.15	1.00	1.02
Latin America	0.83	0.80	0.84	0.76	0.65	0.69	0.81	0.80	0.86	0.93	0.97	0.80	0.84
North America	0.94	0.86	0.88	0.92	0.92	0.92	0.91	0.87	0.87	0.90	0.92	0.87	0.84
EEC	1.11	1.19	1.18	1.19	1.28	1.27	1.14	1.23	1.31	1.25	1.24	1.33	1.24
C.W. Europe	0.78	0.82	0.78	0.76	0.89	1.23	1.06	1.00	1.08	1.14	1.02	1.17	1.16
N.W. Europe	0.78	0.84	0.85	0.84	0.81	0.79	0.81	0.99	1.09	1.09	0.95	1.04	0.98
S.W. Europe	0.73	0.78	0.84	0.88	0.78	0.81	0.79	0.84	0.99	1.04	0.97	1.09	1.07
East. Europe	0.96	0.98	0.99	0.98	0.95	0.95	0.97	1.01	1.01	1.02	1.05	1.06	1.06
USSR b/	1.05	1.05	1.04	1.02	0.99	1.00	1.00	1.01	0.99	0.98	0.99	0.98	0.98
Middle East	0.11	0.10	0.09	0.11	0.11	0.08	0.08	0.18	0.14	0.16	0.14	0.15	0.14

a/   Excluding South Africa.

b/   USSR consumption for the period 1977-1982 has been estimated, as described in Chapter I.

CHAPTER I. METHODOLOGY USED IN THE PREPARATION OF THE STUDY

1.1 Selection of the most representative types and/or aggregate groups of
     steel products which are to be the subject of detailed consideration

The analysis of the trends in international steel trade is based on the
statistical data presented in the UN/ECE publication Statistics of World Trade
in Steel, complemented by external sources with compatible lists of products,
such as COMTRADE (see below).

Statistics of World Trade in Steel is compiled from replies to
questionnaires submitted by countries or from official national sources. It
contains basic data on exports of semi-finished and finished steel products,
which are divided into 12 categories. Ingots and semis are considered
semi-finished products and all others are considered finished. Definitions of
individual commodities are provided in Annex I. Data are given in thousands
of metric tons and in accordance with commonly agreed definitions which have
been worked out by the ECE Steel Committee. Steel trade statistics are
reported to ECE on the basis of the United Nations Standard International
Trade Classification, Revised (SITC).

The second source of statistical information used in the preparation of
the study is COMTRADE - Commodity Trade Statistics - which is compiled
according to the Standard International Trade Classification and based on
statistics provided by Governments. In order to permit comparison of the two
sources of data, the more detailed information in COMTRADE was aggregated to
the level of the 12 product categories of Statistics of World Trade in Steel.

Statistics usually measure steel production in terms of crude steel.
Crude steel, however, is not normally traded on an international basis; steel
products only are submitted to international exchanges. Table 1.1, reproduced
from IISI World Steel Trade in Figures 1984, while not completely conforming
with the United Nations definition of the steel industry, gives a general view
of the relative importance of the different steel product groups in the steel
exports of 14 exporting countries. It demonstrates the dominance of the
following groups of products:

- Sheet, coil and strip;
- Heavy and light bars and sections;
- Tubes; and
- Plates.

In 1982, sheet, coil and strip represented about 35 per cent of total
steel exports, followed by heavy and light bars and sections at 21 per cent,
tubes and fittings at 18 per cent and plates at 9 per cent. Together, these
four categories accounted for 83 per cent of the total. Ingots and
semi-finished steel accounted for approximately 5 per cent, tinplate for
3 per cent and wire rods and wire for approximately 2 per cent.

Taking into account the above and the available statistical information
and in order to simplify the analysis, it has been decided to arrange the
steel products in the following aggregate groups:

- Ingots and semis;
- Long products (heavy and light sections, railway-track material, wire
  rod);
- Flat products (plates, sheets, hoop and strip, tinplate);
- Tubes and fittings; and
- Wire.

Table 1.1.

World steel exports analysed by products
1974, 1979, 1981 and 1982

Products	Millions of tonnes				Percentage of total			
	1974	1979	1981	1982	1974	1979	1981	1982
Ingots and semi-finished products	6.3	5.6	6.1	5.2	5.9	5.2	5.7	5.3
Railway-track material	1.0	1.0	1.2	0.9	0.9	0.9	1.1	0.9
Bars and rods, hot-rolled	12.2	12.5	11.0	11.2	11.3	11.6	10.3	11.5
Angles, shapes and sections	11.0	9.8	10.5	9.0	10.2	9.1	9.9	9.2
Wire rods	6.1	6.3	5.3	4.4	5.7	5.8	5.0	4.5
Hot-rolled strip	2.2	1.9	1.7	1.3	2.0	1.7	1.6	1.3
Plates	13.0	9.4	9.4	8.5	12.1	8.7	8.8	8.7
Hot-rolled sheets and coils	13.9	15.2	13.8	11.8	12.9	14.1	12.9	12.1
Cold-rolled sheets and coils	16.1	16.5	15.0	14.1	15.0	15.3	14.1	14.5
Galvanized sheets	3.4	4.6	4.3	4.2	3.2	4.3	4.0	4.3
Tinplate and black plate	3.4	3.1	3.2	3.1	3.2	2.9	3.0	3.2
Other coated sheets	0.8	1.3	1.3	1.3	0.7	1.2	1.2	1.3
Wheels (rolled and forged) and axles	0.3	0.2	0.2	0.1	0.3	0.2	0.2	0.1
Steel tubes and fittings	12.8	15.5	18.8	18.0	11.9	14.3	17.6	18.5
Drawn wire	2.1	1.9	1.8	1.6	1.9	1.8	1.7	1.6
Cold-rolled strip	2.1	2.2	2.0	1.9	1.9	2.0	1.9	2.0
Cold-finished bars and rods	0.7	0.7	0.7	0.7	0.6	0.6	0.7	0.7
Castings	0.1	0.1	0.1	0.1	0.1	0.1	0.1	0.1
Forgings	0.2	0.2	0.2	0.2	0.2	0.2	0.2	0.2
Total	107.6	107.9	106.5	97.6	100	100	100	100

Source: IISI World Steel Trade in Figures 1984.

Note by IISI: Exports in World Steel Trade in Figures include intra-EEC and intra-CMEA trade and are based on a broad definition of the steel industry and its products, including ingots, semi-finished products, hot-rolled and cold-finished products, tubes and wire and unworked castings and forgings.

1.2 Selection of countries and/or regions whose participation in international steel trade will be analysed

Taking into account the countries and regions whose participation in internationl steel trade is more representative, as well as the steel trade statistical data available, it has been decided to break down the world market for steel as follows:

- European Economic Community;
- USSR;
- Centrally-planned-economy countries of eastern Europe;
- Northern Europe;
- Central Europe;
- Southern Europe;
- Other European countries;
- North America;
- Other North America;
- New industrially developed large steel-producing countries of Latin America;
- Other Latin America;
- Far East, excluding Japan;
- Japan;
- Other Far East;
- Middle East;
- Other Middle East;
- Oceania;
- Other Oceania;
- Africa; and
- Other Africa.

Europe has been divided into the following seven groups of countries:

- European Economic Community of nine (Belgium, Denmark, France, Federal Republic of Germany, Ireland, Italy, Luxembourg, Netherlands, United Kingdom); 1/
- USSR;
- Centrally-planned-economy countries of eastern Europe;
- Northern Europe (Finland, Norway, Sweden);
- Central Europe (Austria, Switzerland);
- Southern Europe (Greece, Portugal, Spain, Turkey, Yugoslavia); and
- Other European countries (all countries of this region which are not included in the other six groups of European countries).

North America has been divided into the following two groups of countries:

- Canada and the United States; and
- Other North America (Bermuda, Greenland, St. Pierre and Miquelon).

Latin America has been divided into the following two groups of countries:

- New industrially developed steel-producing countries of Latin America (Argentina, Brazil, Mexico, Venezuela); and
- Other Latin America (all other countries of this region). 2/

Far East has been divided into the following four groups of countries:

- India;
- Republic of Korea;
- Japan; and
- Other Far East (all other countries of this region). 2/

Middle East has been divided into the following two groups of countries:

- Egypt, Saudi Arabia; and
- Other Middle East (all other countries of this region). 2/

Oceania has been divided into the following two groups of countries:

- Australia and New Zealand; and
- Other Oceania (all other countries of this region). 2/

Africa has been divided into the following two groups of countries:

- North Africa (Algeria, Libyan Arab Jamahiriya, Morocco, Tunisia); and
- Other Africa (except those included in other regions and including South Africa). 2/

The groupings of countries or areas used in the study are a selection of groupings of economic and social interest, used by the United Nations Statistical Office. Some of the groupings comprise countries or areas from various geographical groupings. The countries or areas are those for which statistical data have been or may be compiled by the secretariat.

1.3 Justification of indicators to be used as criteria in examining the different aspects of the evolution of international steel trade

For the purposes of the present study, keeping in mind that the criteria should reflect the evolution of the relative importance of the different exporting and importing countries and regions, the evolution of "geographical and product specialization" of various exporting/importing regions as well as countries, and the evolution of the balance sheets of regions and/or countries participating in international steel trade, the following indicators have been chosen.

The indicators which reflect the changes in regional structure

Changes in the importance of each exporter

$$\frac{X_i}{\sum \sum X_i} \qquad (1)$$

where:

X	-	expresses a tonnage
i	-	designates an exporting country i
X_i	-	export of a country i
$\sum \sum X_i$	-	designates export of all exporting countries

Changes in the importance of each exporting zone

$$\frac{\sum x_i}{\sum \sum x_i} \qquad (2)$$

where:

$\sum x_i$ - designates export of exporting zone i

Changes in the importance of each importer

$$\frac{x_j}{\sum \sum x_j} \qquad (3)$$

where:

   j   - designates an importing country j
   x_j - designates import of a country j
$\sum \sum x_j$ - designates import of all importing countries

Changes in the importance of each importing zone

$$\frac{\sum x_j}{\sum \sum x_j} \qquad (4)$$

where:

$\sum x_j$ - designates import of importing zone j

Changes in balances for countries and regions

$$\begin{array}{c} x_i - x_{ji} \\ \sum x_i - \sum x_{ji} \end{array} \qquad (5)$$

where:

   x_{ji} - designates import of a country i
$\sum x_{ji}$ - designates import of an importing zone i

Changes in the share of each exporter in each importing zone

$$\frac{x_{ij}}{\sum x_j} \qquad (6)$$

where:

　　X_{ij}　　-　designates export of a country i to importing zone j

Changes in the "specialization" of an exporting country i in an importing zone j

$$\frac{X_{ij}}{X_i} \Bigg/ \frac{\sum X_j}{\sum \sum X_j} \qquad (7)$$

where:

$\sum \sum X_j$　-　designates import of all importing countries

The indicators which reflect the changes in the structure of trade, by product

Changes in the share of each product in trade

$$\frac{\sum \sum x^p}{\sum \sum x^t} \qquad (1)$$

where:

　　p　　-　in superscript designates a product
　　t　　-　designates all products
$\sum \sum x^p$　-　designates total export of a product p
$\sum \sum x^t$　-　designates total export of all products

Specialization by product by exporting countries in export

$$\frac{x_i^p}{x_i^t} \qquad (2)$$

where:

　　x_i^p　　-　designates export of a product "p" by country i
　　x_i^t　　-　designates export of all products by country i

Specialization by product by each importing zone in import

$$\frac{\sum x_j^p}{\sum x_j^t} \qquad (3)$$

where:

$\sum x_j^p$　-　designates import of a product "p" by importing zone j
$\sum x_j^t$　-　designates import of all products by importing zone j

The indicator which reflects the specialization of each exporting country in an importing zone, by product

The coefficient of specialization of an exporting country i in an importing zone j for a product P will be calculated for a certain number of products

$$\frac{x_i{}^P{}_j}{x_i{}^t{}_j} \Bigg/ \frac{\sum x_j{}^P}{\sum x_j{}^t} \qquad (4)$$

where:

$x_i{}^P{}_j$ — designates export of a product "p" by a country i to importing zone j

$x_i{}^t{}_j$ — designates export of all products by a country i to importing zone j

$\sum x_j{}^P$ — designates import of product "p" by importing zone j

$\sum x_j{}^t$ — designates import of all products by importing zone j

1.4 The methodology used for estimating steel trade in tonnage terms between the USSR and other centrally-planned-economy countries of eastern Europe for the period from 1977 to 1982

Since there were no statistical data available concerning steel trade in tonnage between the USSR and other centrally-planned economy countries of eastern Europe it was decided to make a variety of assumptions and estimates provided that each of those was supported by reference to its source. The methodology used and results of a method are given below.

The method

The method employed consists in the compilation, in successive stages and with the maximum use of cross-checks, of tables 1.2 and 1.11 to 1.17. The successive stages (1, 2, 3, etc.) indicated in the boxes of those tables for the period 1970-1976, for which the exports of the USSR were available in tonnes, obviously differ from those indicated for the period 1977-1983, for which such data were not directly available.

Period 1970-1976

In the first stage, all the information directly available is entered in table 1.2 (boxes showing the figure 1). This information consists mainly of the figures in the first line concerning the exports of the "four countries" (Bulgaria, Czechoslovakia, Hungary and Poland) and the figures for the total imports of the same four countries (last line). All these figures are of course consolidated, i.e. corrected in the light of trade between the four countries concerned. During this stage, it is also possible to complete the line relating to the exports of the USSR and the total imports and exports of the German Democratic Republic and Romania.

Table 1.2. Stages in estimating steel trade among centrally-planned-economy countries of eastern Europe

(Tonnes)

Period 1970-1976

	Four countries	German Democratic Republic	Romania	Total two countries	USSR	Others	Total
Four countries	\	1	1	1	1	1	1
German Democratic Republic		\	2		1	3	1
Romania		2	\			3	1
Total two countries	2			\	4	3	2
USSR	1	1	1	1	\	1	1
Others	2	2	2	2	2	\	2
Total	1	1	1	2	5		\

Period 1977-1982

	Four Countries	German Democratic Republic	Romania	Total two countries	USSR	Others	Total
Four countries	\	1	1	1	1	1	1
German Democratic Republic		\	5			3	1
Romania		2	\			3	1
Total two countries	5			\	6	3	5
USSR	4	4	4	4	\	4	4
Others	2	2	2	2	2	\	
Total	1	1	1	4	7		\

The second stage is more estimative; it involves an attempt to establish the tonnage exported by countries other than those of eastern Europe to markets in the countries of eastern Europe. This process, which is incomplete in so far as the exercise does not cover all potential exporting countries, makes it possible to fill in the penultimate line of the table (exports of "others"). By calculating the difference between total imports and those shown in the table, it is then possible to estimate the imports of the "four countries" from the German Democratic Republic and Romania, as well as trade between the German Democratic Republic and Romania, which, deducted from the total exports of those two countries, gives the net exports of the group to which they belong.

During the third stage, the COMTRADE data bank is consulted in an attempt to quantify imports from the German Democratic Republic and Romania to the countries covered by that data bank.

Exports from the German Democratic Republic and from Romania to the USSR are calculated simply by deducting exports to the "four countries" and to the "other countries", already estimated during the preceding stages, from total exports (consolidated in the light of trade between the two countries).

Total imports into the USSR can then be calculated simply by adding the figures entered in the column.

This apparently simple and logical method has encountered a number of problems due either to the lack of data for a particular year (especially at the beginning of the period) or to a lack of uniformity among sources.

This is particularly the case with trade between the USSR and Bulgaria. It seems that Bulgarian imports, as defined in Statistics of World Trade in Steel, are not based on the same concept as exports from the USSR to Bulgaria.

Table 1.3. Steel imports of Bulgaria
(Thousands of tonnes)

	1973	1974	1975	1976
Imports by Bulgaria	711	676	720	819
Deliveries from the USSR	719	629	1 553	647
Deliveries to Bulgaria				
from Czechoslovakia, Hungary, Poland	112	121	126	186
Exports from the "others"	285	303	243	201
Difference	-405	-377	-1 202	-215

It also seems that the definitions of Soviet deliveries are not uniform throughout the period (see the difference in 1975).

It may be seen that Bulgaria gives two different figures for its exports and imports depending on whether it is providing information for United Nations statistical bulletins or replying to questionnaires of the Working Party on the Steel Market for "The Steel Market in ..." review, which is published by the Steel Committee of the Economic Commission for Europe.

Table 1.4. Foreign trade of Bulgaria

	The Steel Market in ...		World Trade in Steel		Difference	
	Exports	Imports	Exports	Imports	Exports	Imports
1973	897	1 013	889	718	8	295
1974	887	1 673	810	688	77	985
1975	896	1 957	793	739	103	1 218
1976	1 324	2 192	1 114	849	210	1 343
1977	1 465	2 505	1 049	964	416	1 541
1978	1 514	2 626	1 079	1 131	435	1 475
1979	1 342	2 561	922	1 160	420	1 401
1980	1 343	2 707	1 050	1 385	293	1 322
1981	1 239	2 674	1 073	1 483	166	1 191
1982	1 029	2 541	933	1 378	96	1 163
1983	1 217	2 803	879	1 262	338	1 541

It was not possible to obtain information on the reason for the differences in the above table. In the absence of more detailed information, it was presumed that such differences were due primarily to trade of a non-commercial nature, which must be taken into consideration in order to achieve a certain degree of chronological uniformity for exports from the USSR. Use was therefore made of the import and export figures in "The Steel Market in ...". In the case of exports, the difference (penultimate column in table 1.4) was added to the exports of the "four countries" to the USSR obtained in the first stage on the basis of the data given in the statistics of the World Trade in Steel bulletin. The basic data should be amended accordingly.

Period 1977-1982

The procedure for this period is longer and involves greater uncertainty in so far as it is necessary to estimate the exports of the USSR (stage 4), although, in the other stages, the logic remains the same as that applied to the preceding period. The export tonnage of the USSR is estimated in four stages:

1. Estimate of USSR exports to the German Democratic Republic;

2. Estimate of USSR exports to the group consisting of Bulgaria, Hungary, Czechoslovakia, Poland, Romania and Yugoslavia on the basis of Yugoslav data;

3.   Cross-checking of the above results by country of destination
included in the group considered in stage 2; and

4.   Estimate of USSR exports to the "other countries" (penultimate
column in the tables).

Estimate of USSR exports to the German Democratic Republic

The imports of the German Democratic Republic (excluding imports from the
Federal Republic of Germany) are published annually. Its exports to countries
participating in the COMTRADE data bank make it possible to distinguish the
"four countries" (Bulgaria, Czechoslovakia, Hungary and Poland) from the other
countries. The difference between total imports and the flows recorded by
COMTRADE from 1970 to 1976 can be broken down between the USSR and other
non-COMTRADE countries. The problem is to obtain that breakdown for the years
1977 to 1982. There is a close ratio ($R^2 = 0.86$) between the share of
tonnage imported from non-COMTRADE countries and the level of total imports.
The application of this ratio made it possible to obtain the estimates given
in brackets in table 1.5.

Table 1.5. Imports of the German Democratic Republic
(excluding imports from the Federal Republic of Germany)

	Total imports	COMTRADE countries		Non-COMTRADE countries		
		"Four countries"	Others	USSR	Others	Total
1970	3 490	298	195	2 621	376	2 997
1971	3 260	352	116	2 578	214	2 792
1972	3 637	424	222	2 582	409	2 991
1973	3 736	397	149	2 645	545	3 190
1974	4 078	407	179	2 762	730	3 492
1975	4 004	462	178	2 888	475	3 364
1976	4 075	470	169	2 953	483	3 436
1977	4 159	447	141	(3 014)	(557)	3 571
1978	4 210	536	113	(3 011)	(550)	3 561
1979	4 241	475	136	(3 100)	(530)	3 630
1980	4 263	504	127	(3 122)	(510)	3 632
1981	4 493	454	103	(3 266)	(670)	3 936
1982	4 574	459	108	(3 352)	(655)	4 007

Estimate of USSR exports to the six countries: Bulgaria, Czechoslovakia,
Hungary, Poland, Romania and Yugoslavia

For the period 1970 to 1976, a ratio has been calculated between:

PARVAL - the share of each country in exports, by value, from the USSR to
the group of six countries in question; and

PARTON - the share of each country in exports, by tonnage, from the USSR to
the group of six countries in question.

This ratio takes the following form:

$$PARTON = a \ PARVAL + bt + c \quad (t = time)$$

The value of the parameters, their significance (Student's t distribution in
brackets) and the results of the usual tests are shown in table 1.6.

Table 1.6. Value of parameters, their significance and results of tests

	a	b	c	R^2	DW
Bulgaria	1.35 (9.9)	0.62 (1.7)	-12.41	0.96	2.56
Czechoslovakia	0.64 (7.9)	-0.33 (1.3)	3.74	0.95	2.64
Hungary	0.93 (2.5)	-0.20 (-0.5)	4.07	0.72	2.12
Poland	0.71 (5.3)	-0.27 (-1.6)	5.66	0.91	3.10
Romania	1.34 (7.1)	-0.02 (-1.2)	-0.82	0.96	1.76
Yugoslavia	0.61 (3.1)	-0.38 (-1.-)	4.94	0.82	2.15

The total, to which the PARTON percentages obtained on the basis of the
1977-1982 PARVAL values are applied, is itself obtained by using the Yugoslav
data, which are available in terms both of tonnage and of value.

$$\frac{\text{Yugoslav imports in tonnes}}{\text{PARTON}} = \text{USSR exports to the six countries}$$

The tonnages thus obtained were regarded merely as a starting point for a
first iteration, which was subsequently corrected by cross-checking for each
of the countries concerned. In the case of Poland, data on imports from the
USSR were available for the years 1980-1981 and 1982; such real data were, of
course, retained.

Estimate of USSR exports to the other countries

The data available for the period 1970-1976 make it possible to distinguish between two major groups of countries of destination:

The other countries specified;

A non-specified group (unallocated).

For the period 1977-1982, the first group can be identified on the basis of imports from the USSR to the countries covered by the COMTRADE data bank. However, there are several years in which exports to Yugoslavia were not taken into account and no significant tonnage was recorded for western Europe in 1982. The following corrections were therefore made on the basis of Yugoslav and EEC sources.

Table 1.7.   Tonnages to be added to USSR exports recorded by COMTRADE

	Recorded	Yugoslavia	EEC	Total
1977	862	-	-	862
1978	197	273	-	470
1979	291	-	-	291
1980	367	297	-	664
1981	143	304	-	447
1982	-	308	150	458

In the case of the second group, which seems to consist of destinations such as Cuba, Viet Nam and Mongolia, its share of USSR exports, by value, is known and a chronological modulation can therefore be obtained.

Table 1.8.   Chronological modulation

	Total	Latin America	Far East
1977	500	385	115
1978	495	400	95
1979	570	425	145
1980	470	360	110
1981	560	395	165
1982	685	430	255

Table 1.9.   USSR exports to other countries

	Other countries specified	Unallocated tonnage	Percentage of total exports by value
1970	999	477	
1971	853	370	
1972	1 489	509	
1973	1 303	430	
1974	1 169	470	
1975	1 010	578	
1976	1 478	74	
1977	862	(500)	7.0
1978	470	(495)	7.4
1979	291	(570)	8.3
1980	664	(470)	7.2
1981	447	(560)	8.9
1982	458	(685)	10.5

Results

The detailed results obtained for each year are shown in the above tables. As can be seen, the calculations could be made only with effect from 1973. For earlier years (1970-1972), some data are missing, particularly on the foreign trade of Romania. If such data became available, the tables could easily be completed.

The results have been summarized in table 1.10, which shows the exports, imports and balance obtained for the USSR on an overall basis, and in relation to the other centrally-planned-economy countries of eastern Europe.

The estimated results have been summarized in table 1.10, which shows the exports, imports and balance obtained for the USSR on an overall basis and in relation to the other centrally-planned-economy countries of eastern Europe.

Since there are no official statistical data (in tonnage terms) available on USSR foreign trade in steel as of 1977, the data cited in table 1.10 have been calculated according to the above methodology. The data are therefore approximate.

These estimated data are also used in all the other tables and comments contained in the present study.

Table 1.10. USSR foreign trade

	Total			Other countries of eastern Europe		
	Exports	Imports	Balance	Exports	Imports	Balance
1973	7 050	6 031	1 019	5 317	2 041	3 276
1974	6 889	7 652	- 763	5 170	1 713	3 457
1975	7 825	7 326	499	6 189	1 773	4 416
1976	7 503	9 532	-2 029	6 011	1 826	4 185
1977	7 387	7 326	61	6 025	1 946	4 079
1978	7 368	8 932	-1 564	6 336	2 029	4 307
1979	7 407	9 409	-2 002	6 546	2 061	4 485
1980	7 184	9 064	-1 880	6 050	2 152	3 898
1981	7 089	8 921	-1 832	6 082	2 090	3 992
1982	7 575	10 083	-2 508	6 331	2 586	3 745

Table 1.11. Centrally-planned-economy countries of eastern Europe:
trade in steel products, 1970 and 1971

(Tonnes)

1970	Four countries	German Democratic Republic	Romania	Total two countries	USSR	Others	Total
Four countries		298	345	643	715	3 691	5 049
German Democratic Republic						124	
Romania		48				470	
Total two countries						594	
USSR	2 337	2 621	603	3 224		1 516	7 077
Others	743	523	485	1 008	1 668		
Total		3 490					

1971	Four countries	German Democratic Republic	Romania	Total two countries	USSR	Others	Total
Four countries		352	282	634	610	4 209	5 453
German Democratic Republic						154	
Romania		40				571	
Total two countries						725	
USSR	2 298	2 578	661	3 239		1 264	6 801
Others	924	290	533	823	2 050		
Total		3 260					

Table 1.12. Centrally-planned-economy countries of eastern Europe:
trade in steel products, 1972 and 1973
(Tonnes)

1972	Four countries	German Democratic Republic	Romania	Total two countries	USSR	Others	Total
Four countries		424	284	708	716	4 534	5 958
German Democratic Republic						203	
Romania		70				583	
Total two countries						786	
USSR	2 129	2 582	560	3 142		2 039	7 310
Others	994	561	447	1 008	2 736		
Total		3 637					

1973	Four countries	German Democratic Republic	Romania	Total two countries	USSR	Others	Total
Four countries		397	319	716	592	4 184	5 492
German Democratic Republic			156			252	1 496
Romania		80				453	1 388
Total two countries	494				1 449	705	2 648
USSR	2 129	2 645	543	3 188		1 733	7 050
Others	1 949	614	475	1 089	3 990		
Total	4 572	3 736	1 493	4 993	6 031		

Table 1.13. Centrally-planned-economy countries of eastern Europe:
trade in steel products, 1974 and 1975

(Tonnes)

1974	Four countries	German Democratic Republic	Romania	Total two countries	USSR	Others	Total
Four countries		407	309	716	881	3 652	5 249
German Democratic Republic			130			274	1 836
Romania		90				333	1 308
Total two countries	1 485				832	607	2 924
USSR	1 888	2 762	520	3 282		1 719	6 889
Others	2 628	819	552	1 371	5 939		
Total	6 001	4 078	1 511	5 369	7 652		

1975	Four countries	German Democratic Republic	Romania	Total two countries	USSR	Others	Total
Four countries		462	554	1 016	892	4 697	6 605
German Democratic Republic			426			296	1 517
Romania		80				422	1 477
Total two countries	889				881	718	2 488
USSR	2 967	2 864	358	3 222		1 636	7 825
Others	2 608	573	512	1 085	5 553		
Total	6 464	4 004	1 850	5 323	7 326		

Table 1.14. Centrally-planned-economy countries of eastern Europe: trade in steel products, 1976 and 1977
(Tonnes)

1976	Four countries	German Democratic Republic	Romania	Total two countries	USSR	Others	Total
Four countries		470	559	1 029	1 026	4 651	6 706
German Democratic Republic			260			535	1 491
Romania		80				625	1 675
Total two countries	1 898				800	1 160	2 826
USSR	2 538	2 953	520	3 473		1 492	7 503
Others	2 692	572	693	1 265	7 706		
Total	7 128	4 075	2 032	5 767	9 532		

1977	Four countries	German Democratic Republic	Romania	Total two countries	USSR	Others	Total
Four countries		447	553	1 000	1 096	5 569	7 665
German Democratic Republic			172			568	1 654
Romania		85				805	1 899
Total two countries	1 073				850	1 373	3 296
USSR	2 411	3 014	600	3 614		1 362	7 387
Others	1 651	613	444	1 057	5 380		
Total	5 921	4 159	1 769	5 671	7 326		

Table 1.15. Centrally-planned-economy countries of eastern Europe: trade in steel products, 1978 and 1979

(Tonnes)

1978	Four countries	German Democratic Republic	Romania	Total two countries	USSR	Others	Total
Four countries		536	443	979	1 276	5 139	7 394
German Democratic Republic			607			550	1 580
Romania		90				989	2 240
Total two countries	831				753	1 539	3 123
USSR	2 869	3 011	523	3 534		965	7 368
Others	1 480	573	557	1 130	6 903		
Total	5 913	4 210	2 130	5 643	8 932		

1979	Four countries	German Democratic Republic	Romania	Total two countries	USSR	Others	Total
Four countries		475	410	885	1 251	5 308	7 444
German Democratic Republic			300			541	1 785
Romania		90				934	1 948
Total two countries	1 058				810	1 475	3 343
USSR	3 151	3 100	295	3 395		861	7 407
Others	1 399	576	668	1 244	7 348		
Total	5 608	4 241	1 673	5 524	9 409		

Table 1.16. Centrally-planned-economy countries of eastern Europe: trade in steel products, 1980 and 1981
(Tonnes)

1980	Four countries	German Democratic Republic	Romania	Total two countries	USSR	Others	Total
Four countries		504	292	796	1 152	5 296	7 244
German Democratic Republic			174			537	2 153
Romania		95				728	1 984
Total two countries	1 603				1 000	1 265	3 868
USSR	2 598	3 122	330	3 452		1 134	7 184
Others	962	542	390	932	6 912		
Total	5 163	4 263	1 274	5 180	9 064		

1981	Four countries	German Democratic Republic	Romania	Total two countries	USSR	Others	Total
Four countries		454	268	722	1 090	4 852	6 664
German Democratic Republic			176			673	2 584
Romania		100				1 074	2 083
Total two countries	1 644				1 000	1 747	4 391
USSR	2 511	3 266	305	3 571		1 007	7 089
Others	679	673	225	898	6 831		
Total	4 834	4 493	974	5 191	8 921		

Table 1.17. Centrally-planned-economy countries of eastern Europe:
trade in steel products, 1982
(Tonnes)

1982	Four countries	German Democratic Republic	Romania	Total two countries	USSR	Others	Total
Four countries		459	214	673	1 115	4 406	6 274
German Democratic Republic			197			701	2 498
Romania		110				606	1 901
Total two countries	1 314				1 471	1 307	4 092
USSR	2 740	3 352	340	3 692		1 143	7 575
Others	544	653	164	817	7 497		
Total	4 598	4 574	915	5 182	10 083		

CHAPTER II. ANALYSIS OF STRUCTURAL CHANGES IN THE EXPORTS OF STEEL
PRODUCTS FROM VARIOUS COUNTRIES AND/OR REGIONS

2.1 Analysis of changes in geographic orientation of total export deliveries
of steel products from various countries and/or regions

Tables 2.1 and 2.2 summarize the geographic pattern of world steel
exports, excluding and including intra-regional trade, and show the major
exporting regions and their relative positions as exporters.

The data presented in tables 2.3 and 2.4 give a picture of the
contribution made by the principal exporting regions or countries to the total
international supply.

The growth of trade has been accompanied by some remarkable shifts in the
geographical orientation of total export deliveries of steel products. One of
the very important developments during the period has been the emergence of
new exporters from developing countries which are competing with traditional
steel traders, at both the domestic and international levels. Amongst the new
steel exporters are the Republic of Korea, Argentina, Brazil, Mexico and
Venezuela. These shifts in trade and competitiveness have in turn led to a
growing interdependence of markets.

In this context, the following regions should be mentioned:

- The Far East increased its exports during the period from 0.99 million
  to 6.78 million tonnes and its share in world steel exports grew from
  1.59 to 6.84 per cent.

- Latin America increased its exports from 0.54 million to 3.13 million
  tonnes and its share more than tripled, rising from 0.87 per cent to
  3.15 per cent; and

- Southern Europe increased its exports by a factor of almost 10 while
  its share grew from 0.99 to 6.18 per cent.

Japan and EEC retained their leading positions in world steel exports.
Japan increased its exports significantly, from 17.47 million to
28.61 million tonnes and its share in world steel exports fluctuated
between 28 and 39 per cent. EEC also increased its exports considerably, from
18.99 million to 25.83 million tonnes, and its share in world steel exports
varied between 25 per cent and 35 per cent.

One of the major steel-producing regions, North America, recorded a
decreasing trend in its steel exports. During the period, its exports
decreased from 6.12 million to 2.94 million tonnes and its share accordingly
from 9.8 to 2.97 per cent.

The volume of the USSR exports remained relatively stable at around
7.5 million tonnes but its share in world exports of steel products declined
from 11.86 per cent in 1970 to 7.64 per cent in 1982.

Table 2.1. Exports of all products by region from 1970 to 1982 a/

(Thousands of tonnes)

Exporters	1970	1971	1972	1973	1974	1975	1976	1977	1978	1979	1980	1981	1982
Africa	224	100	281	402	343	196	818	1636	1292	1236	1151	768	1075
Far-East	992	730	1341	1086	1507	1370	2723	2418	2877	4059	4377	5690	6783
Japan	17471	23076	20814	24707	32103	28834	35883	33578	30876	30617	29631	28416	28606
Oceania	801	320	713	1211	1033	1480	2047	2334	2507	2124	1504	1048	1262
North America	6120	2112	2215	3203	4251	2321	2472	1590	2186	2418	4685	2638	2944
Latin America	545	560	784	632	299	139	541	503	1424	1557	1693	2207	3126
EEC (9)	18992	22510	23634	26451	32936	27615	22526	27498	33335	31641	28673	32620	25831
N.W. Europe	1838	1863	2136	2332	2281	1997	2139	2676	3277	3383	2906	3023	3073
S.W. Europe	621	1265	1945	2482	1547	1961	3242	3008	4732	4880	5018	5665	6127
C.W. Europe	1304	1361	1404	1387	1749	2053	2208	2130	2463	2912	2768	3104	2889
Eastern Europe	6133	6832	7727	7786	7549	7490	8997	9738	9733	9794	10207	10373	9819
USSR	7409	7373	7327	7028	6839	7773	7502	7387	7368	7407	7184	7089	7575
World	62450	68043	70322	78708	92436	83229	91099	94497	102069	102028	99796	102661	99110

a/ Excluding intra-regional trade.

Table 2.2 Exports of all products by region from 1970 to 1982 b/

(Thousands of tonnes)

Exporters	1970	1971	1972	1973	1974	1975	1976	1977	1978	1979	1980	1981	1982
Africa	281	150	332	452	407	232	875	1682	1363	1295	1203	797	1086
Far East	1123	860	1430	1284	1683	1580	2978	2950	3424	4952	5556	7152	8313
Japan	17471	23076	20814	24707	32103	28834	35883	33578	30876	30617	29631	28416	28606
Oceania	1057	557	865	1424	1272	1745	3313	2546	2711	2475	1758	1231	1505
North America	7759	3896	3880	5016	6869	3830	4011	3570	5091	5292	7363	6337	5020
Latin America	1168	963	1378	1118	746	249	759	910	1994	2364	2336	3103	3771
EEC (9)	41737	45638	50147	54746	63068	51252	50096	52930	60084	61149	58203	60366	50279
N.W. Europe	2340	2348	2763	3084	3064	2707	2847	3315	4012	4133	3730	3795	3761
S.W. Europe	762	1297	1994	2718	1888	2243	3353	3204	4919	5131	5351	5964	6373
C.W. Europe	1472	1521	1612	1572	1955	2210	2357	2271	2625	3097	2947	3264	3046
Eastern Europe	7367	8026	9207	9359	9320	9496	10849	11588	11504	11492	11731	11816	11232
USSR	7409	7373	7327	7028	6839	7773	7502	7387	7368	7407	7184	7089	7575
World	89948	95705	101748	112510	129215	112152	124823	125931	135971	139404	136993	139333	130566

b/ Including intra-regional trade.

Table 2.3. Exports of all products by region from 1970 to 1982 a/

(Percentage)

Exporters	1970	1971	1972	1973	1974	1975	1976	1977	1978	1979	1980	1981	1982
Africa	0.36	0.15	0.40	0.51	0.37	0.24	0.90	1.73	1.27	1.21	1.15	0.75	1.08
Far East	1.59	1.07	1.91	1.38	1.63	1.65	2.99	2.56	2.82	3.98	4.39	5.54	6.84
Japan	27.98	33.91	29.60	31.39	34.73	34.64	39.39	35.53	30.25	30.01	29.69	27.68	28.86
Oceania	1.28	0.47	1.01	1.54	1.12	1.78	2.25	2.47	2.46	2.08	1.51	1.02	1.27
North America	9.80	3.10	3.15	4.07	4.60	2.79	2.71	1.68	2.14	2.57	4.69	2.57	2.97
Latin America	0.87	0.82	1.11	0.80	0.32	0.17	0.59	0.53	1.40	1.53	1.70	2.15	3.15
EEC (9)	30.41	33.08	33.61	33.61	35.63	33.18	24.73	29.10	32.66	31.01	28.73	31.77	26.06
N.W. Europe	2.94	2.74	3.04	2.96	2.47	2.40	2.35	2.83	3.21	3.32	2.91	2.95	3.10
S.W. Europe	0.99	1.77	2.77	3.15	1.67	2.36	3.56	3.18	4.64	4.78	5.03	5.54	6.18
C.W. Europe	2.09	2.00	2.00	1.76	1.89	2.47	2.42	2.25	2.41	2.85	2.77	3.02	2.92
Eastern Europe	9.82	10.04	10.99	9.89	8.17	9.00	9.88	10.30	9.54	9.60	10.23	10.10	9.91
USSR	11.86	10.84	10.42	8.93	7.40	9.34	8.24	7.82	7.22	7.26	7.20	6.91	7.64
World	100	100	100	100	100	100	100	100	100	100	100	100	100

a/ Excluding intra-regional trade.

Table 2.4. Exports of all products by region from 1970 to 1982 b/

(Percentage)

Exporters	1970	1971	1972	1973	1974	1975	1976	1977	1978	1979	1980	1981	1982
Africa	0.31	0.16	0.33	0.40	0.31	0.21	0.70	1.34	1.00	0.93	0.88	0.57	0.83
Far East	1.25	0.90	1.41	1.14	1.30	1.41	2.39	2.34	2.52	3.55	4.06	5.13	6.37
Japan	19.42	24.11	20.46	21.96	24.84	25.71	28.75	26.66	22.71	21.96	21.63	20.39	21.91
Oceania	1.18	0.58	0.85	1.27	0.98	1.56	2.65	2.02	1.99	1.78	1.28	0.88	1.15
North America	8.63	4.07	3.81	4.46	5.32	3.41	3.21	2.84	3.74	3.80	5.38	4.55	3.84
Latin America	1.30	1.01	1.35	0.99	0.58	0.22	0.61	0.72	1.47	1.70	1.70	2.23	2.89
EEC (9)	46.40	47.69	49.29	48.66	48.81	45.70	40.13	42.03	44.19	43.86	42.49	43.33	38.51
N.W. Europe	2.60	2.45	2.72	2.74	2.37	2.41	2.28	2.63	2.95	2.96	2.72	2.72	2.88
S.W. Europe	0.85	1.36	1.96	2.42	1.46	2.00	2.69	2.54	3.62	3.68	3.91	4.28	4.88
C.W. Europe	1.64	1.59	1.58	1.40	1.51	1.97	1.89	1.80	1.93	2.22	2.15	2.34	2.33
Eastern Europe	8.19	8.39	9.05	8.32	7.21	8.47	8.69	9.20	8.46	8.24	8.56	8.48	8.60
USSR	8.24	7.70	7.20	6.25	5.29	6.93	6.01	5.87	5.42	5.31	5.24	5.09	5.80
World	100	100	100	100	100	100	100	100	100	100	100	100	100

b/ Including intra-regional trade.

Central Europe and northern Europe increased their exports by 221 and 167 per cent, respectively, and their share in world steel exports increased by 140 per cent and 105 per cent, respectively.

Oceania increased its exports from 0.80 million to 1.26 million tonnes whilst its share remained virtually unchanged, dropping from 1.28 to 1.27 per cent.

The volume of exports from eastern Europe increased from 6.13 million to 9.82 million tonnes and its share in total world steel exports was stable, at about 9.9 per cent.

The data shows that the large producing countries together with eastern Europe accounted for an overwhelming proportion of total steel exports (75 per cent) in 1982.

Thus, world steel exports are heavily concentrated on a small number of exporting countries. Japan and EEC occupy a special position among the main exporting countries.

Africa remained the smallest exporting region. Its exports increased from 0.22 million to 1.08 million tonnes whilst its share of world steel exports attained slightly more than 1 per cent in 1982.

Changes in the geographic orientation of EEC exports

EEC exports of steel products grew between 1970 (18.99 million tonnes) and 1974 (32.94 million tonnes). Since then they have fluctuated. In 1976, they fell to their lowest level (22.53 million tonnes) since 1971. There was an increase in 1978, when exports grew to 33.34 million tonnes. In 1982, they amounted to 25.83 million tonnes.

The changes in the geographic orientation of EEC exports of steel products during the period from 1970 to 1982 (for selected years) are presented in table 2.5.

While EEC exported steel products to all regions, changes have occurred in the orientation of its exports.

The major recipient region of EEC exports was North America but the volume of EEC exports to this region decreased from 5.21 million tonnes in 1970 to 4.8 million tonnes in 1982, after a peak of 6.8 million tonnes in 1978. The share of EEC exports to North America declined from 27.5 to 18.6 per cent.

The volume of EEC exports to the USSR increased considerably, from 0.96 million to 4.2 million tonnes, the share growing from 5 to 16.2 per cent.

Table 2.5. Changes in the geographic orientation of exports
of steel products

Years / Importers	European Economic Community								North European countries							
	1970		1974		1978		1982		1970		1974		1978		1982	
	1000t	%	1000t	%	1000t	%	1000t	%	1000t	%	1000t	%	1000t	%	1000t	%
Africa	1929	10.2	3466	10.5	2989	9.0	2379	9.2	15	0.8	48	2.1	78	2.4	51	1.7
Algeria	381	2.0	759	2.3	969	2.9	673	2.6	0	0.0	8	0.3	66	2.0	17	0.5
Liby.Arab Jam.	128	0.7	514	1.6	228	0.7	195	0.8	0	0.0	6	0.3	0	0.0	14	0.5
Morocco	176	0.9	301	0.9	269	0.8	235	0.9	5	0.3	1	0.0	0	0.0	0	0.0
Tunisia	49	0.3	99	0.3	129	0.4	192	0.7	0	0.0	0	0.0	0	0.0	5	0.2
Above listed	735	3.9	1673	5.1	1594	4.8	1295	5.0	5	0.3	15	0.7	67	2.0	36	1.2
South Africa	150	0.8	580	1.8	68	0.2	70	0.3	5	0.3	14	0.6	1	0.0	5	0.2
Other Africa	1043	5.5	1213	3.7	1328	4.0	1014	3.9	5	0.3	19	0.8	10	0.3	10	0.3
Far East	1215	6.4	1592	4.8	4287	12.9	2174	8.4	60	3.2	41	1.8	174	5.3	103	3.3
China	420	2.2	520	1.6	2514	7.5	456	1.8	37	2.0	12	0.5	33	1.0	10	0.3
India	221	1.2	366	1.1	630	1.9	803	3.1	5	0.2	12	0.5	11	0.4	43	1.4
Korea,Rep.of	4	0.0	7	0.0	247	0.7	29	0.1	0	0.0	0	0.0	2	0.1	2	0.1
Above listed	644	3.4	894	2.7	3391	10.2	1288	5.0	42	2.3	24	1.1	46	1.4	56	1.8
Japan	5	0.0	11	0.0	10	0.0	8	0.0	4	0.2	5	0.2	27	0.8	3	0.1
Other Far East	565	3.0	687	2.1	886	2.7	878	3.4	14	0.8	13	0.6	101	3.1	44	1.4
Middle East	1010	5.3	2768	8.4	4175	12.5	2609	10.1	12	0.7	36	1.6	119	3.6	130	4.2
Egypt	64	0.3	207	0.6	162	0.5	274	1.1	0	0.0	2	0.1	2	0.0	1	0.0
Saudi Arabia	65	0.3	254	0.8	694	2.1	591	2.3	0	0.0	0	0.0	7	0.2	64	2.1
Above listed	129	0.7	462	1.4	856	2.6	865	3.3	0	0.0	2	0.1	9	0.3	65	2.1
Other Middle East	880	4.6	2306	7.0	3319	10.0	1745	6.8	12	0.7	34	1.5	111	3.4	65	2.1
Oceania	210	1.1	182	0.6	97	0.3	129	0.5	10	0.5	12	0.5	7	0.2	7	0.2
Australia	59	0.3	74	0.2	46	0.1	74	0.3	9	0.5	12	0.5	7	0.2	6	0.2
New Zealand	99	0.5	79	0.2	26	0.1	33	0.1	0	0.0	0	0.0	0	0.0	0	0.0
Other Oceania	52	0.3	30	0.1	24	0.1	22	0.1	0	0.0	0	0.0	-	-	-	-
North America	5214	27.5	6649	20.2	6786	20.4	4798	18.6	87	4.7	158	6.9	433	13.2	418	13.6
Canada	317	1.7	731	2.2	454	1.4	316	1.2	15	0.8	24	1.1	22	0.7	21	0.7
United States	4892	25.8	5911	17.9	6328	19.0	4462	17.3	72	3.9	134	5.9	411	12.5	397	12.9
N. America unalloc.	5	0.0	6	0.0	4	0.0	19	0.1	0	0.0	0	0.0	-	-	0	0.0
Other America	1152	6.1	2958	9.0	2476	7.4	1622	6.3	29	1.6	74	3.2	89	2.7	33	1.1
Argentina	228	1.2	257	0.8	199	0.6	190	0.7	7	0.4	8	0.4	3	0.1	15	0.5
Brazil	177	0.9	1405	4.3	229	0.7	196	0.8	11	0.6	47	2.1	8	0.2	4	0.1
Mexico	23	0.1	160	0.5	924	2.8	430	1.7	4	0.2	9	0.4	8	0.3	7	0.2
Venezuela	106	0.6	360	1.1	604	1.8	353	1.4	3	0.1	2	0.1	3	0.1	2	0.1
Above listed	534	2.8	2183	6.6	1956	5.9	1168	4.5	25	1.3	67	2.9	22	0.7	28	0.9
Oth. Oth. America	617	3.3	775	2.4	520	1.6	454	1.8	4	0.2	7	0.3	67	2.1	5	0.2
Western Europe	6509	34.3	8711	26.4	6476	19.4	7478	28.9	1537	83.6	1718	75.3	2183	66.6	2215	72.1
B.L.E.U.	.	.	.	.	.	.	.	.	39	2.1	71	3.1	110	3.3	63	2.0
Denmark	.	.	.	.	.	.	.	.	344	18.7	331	14.5	276	8.4	374	12.2
France	.	.	.	.	.	.	.	.	119	6.5	203	8.9	207	6.3	183	6.0
Germany,Fed.Rep.	.	.	.	.	.	.	.	.	473	25.7	414	18.2	639	19.5	645	21.0
Ireland	.	.	.	.	.	.	.	.	3	0.2	5	0.2	10	0.3	15	0.5
Italy	.	.	.	.	.	.	.	.	78	4.2	56	2.5	41	1.2	75	2.4
Netherlands	.	.	.	.	.	.	.	.	114	6.2	179	7.8	251	7.7	235	7.6
United Kingdom	.	.	.	.	.	.	.	.	251	13.7	308	13.5	551	16.8	404	13.2
EEC(9)	.	.	.	.	.	.	.	.	1421	77.3	1567	68.7	2085	63.6	1994	64.9
Finland	443	2.3	460	1.4	170	0.5	248	1.0	.	.	.	.	.	.	.	.
Iceland	14	0.1	29	0.1	22	0.1	21	0.1	.	.	.	.	.	.	.	.
Norway	826	4.4	1106	3.4	697	2.1	614	2.4	.	.	.	.	.	.	.	.
Sweden	1237	6.5	1635	5.0	1092	3.3	1322	5.1	.	.	.	.	.	.	.	.
Northern Europe	2521	13.3	3230	9.8	1980	5.9	2205	8.5	.	.	.	.	.	.	.	.
Greece	359	1.9	486	1.5	719	2.2	691	2.7	3	0.2	5	0.2	2	0.1	17	0.6
Portugal	274	1.4	577	1.8	446	1.3	523	2.0	35	1.9	28	1.2	7	0.2	40	1.3
Spain	907	4.8	1059	3.2	673	2.0	1550	6.0	29	1.6	64	2.8	23	0.7	80	2.6
Turkey	141	0.7	797	2.4	286	0.9	249	1.0	1	0.0	5	0.2	2	0.1	2	0.1
Yugoslavia	326	1.7	462	1.4	324	1.0	340	1.3	7	0.4	9	0.4	8	0.2	28	0.9
Southern Europe	2006	10.6	3380	10.3	2447	7.3	3352	13.0	75	4.1	111	4.9	42	1.3	167	5.4
Austria	272	1.4	439	1.3	653	2.0	605	2.3	15	0.8	18	0.8	19	0.6	22	0.7
Switzerland	1635	8.6	1580	4.8	1312	3.9	1259	4.9	27	1.5	21	0.9	36	1.1	31	1.1
Central Europe	1907	10.0	2019	6.1	1966	5.9	1864	7.2	42	2.3	39	1.7	54	1.7	53	1.7
W.Europe unalloc	75	0.4	82	0.2	83	0.2	57	0.2	0	0.0	1	0.1	1	0.0	1	0.0
Eastern Europe	1740	9.2	6570	19.9	6040	18.1	4634	17.9	87	4.7	193	8.5	194	5.9	116	3.8
Albania	13	0.1	24	0.1	9	0.0	14	0.1	0	0.0	0	0.0	0	0.0	1	0.0
Bulgaria	74	0.4	232	0.7	248	0.7	135	0.5	1	0.0	1	0.0	4	0.1	9	0.3
Czechoslovakia	131	0.7	60	0.2	169	0.5	33	0.1	2	0.1	3	0.1	2	0.1	2	0.1
German Dem.Rep.	18	0.1	42	0.1	44	0.1	32	0.1	25	1.4	27	1.2	19	0.6	7	0.2
Hungary	89	0.5	137	0.4	122	0.4	67	0.3	2	0.1	2	0.1	11	0.3	10	0.3
Poland	229	1.2	1598	4.9	599	1.8	55	0.2	11	0.6	114	5.0	41	1.2	11	0.4
Romania	228	1.2	329	1.0	384	1.2	102	0.4	4	0.2	12	0.5	10	0.3	2	0.1
Above listed	782	4.1	2421	7.4	1574	4.7	438	1.7	44	2.4	158	6.9	87	2.7	41	1.3
USSR	958	5.0	4149	12.6	4466	13.4	4196	16.2	43	2.3	34	1.5	106	3.3	75	2.4
E.Europe unalloc	0	0.0	0	0.0	-	-	-	-	0	0.0	-	-	-	-	-	-
Unallocated	15	0.1	41	0.1	8	0.0	6	0.0	0	0.0	0	0.0	-	-	-	-
World	18992	100.0	32936	100.0	33335	100.0	25831	100.0	1838	100.0	2281	100.0	3277	100.0	3073	100.0

EEC increased both the volume and the share of its exports to the following regions:

- Far East: Exports - from 1.21 million to 2.17 million tonnes. Share - from 6.4 to 8.4 per cent;

- Middle East: Exports - from 1.01 million to 2.61 million tonnes. Share - from 5.3 to 10.1 per cent;

- Latin America (four countries): Exports - from 0.53 million to 1.17 million tonnes. Share - from 2.8 to 4.5 per cent; and

- Southern Europe: Exports - from 2.01 million to 3.35 million tonnes. Share - from 10.6 to 13 per cent.

EEC showed a decrease in both the volume and the share of its exports to the following regions:

- Oceania: Exports - from 0.21 million to 0.13 million tonnes. Share - from 1.1 to 0.5 per cent;

- Northern Europe: Exports - from 2.52 million to 2.21 million tonnes. Share - from 13.3 to 8.5 per cent;

- Central Europe: Exports - from 1.91 million to 1.86 million tonnes. Share - from 10.0 to 7.2 per cent; and

- Eastern Europe: Exports - from 0.78 million to 0.44 million tonnes. Share - from 4.1 to 1.7 per cent.

Although EEC exports to Africa increased from 1.93 million to 2.38 million tonnes, the share of those exports declined from 10.2 to 9.2 per cent.

Changes in the geographic orientation of Japan's exports

Japan is the world's largest exporter of finished and semi-finished steel products. Its exports increased from 17.47 million to 35.88 million tonnes between 1970 and 1976. However, starting in 1977, a decreasing trend became evident in the volume of exports of steel from Japan (table 2.1). Between 1977 and 1982, Japanese direct exports decreased from 33.58 million to 28.61 million tonnes, albeit with some fluctuations.

The changes which occurred in the geographic orientation of Japanese exports between 1970 and 1982 (for selected years) are presented in table 2.6.

As can be seen, Japan was firmly established as the principal supplier of foreign produced steel products to the Far East and the main part of Japanese exports was directed to this region. The volume of its exports increased from 5.25 million to 12 million tonnes. Its share in total Japanese exports increased from 30.1 per cent in 1970 to 41.9 per cent in 1982.

Table 2.6. Changes in the geographic orientation of exports of steel products

	Japan								Far East							
Years	1970		1974		1978		1982		1970		1974		1978		1982	
Importers	1000t	%	1000t	%	1000t	%	1000t	%	1000t	%	1000t	%	1000t	%	1000t	%
Africa	792	4.5	1353	4.2	1054	3.4	931	3.3	35	3.5	29	1.9	58	2.0	80	1.2
Algeria	75	0.4	237	0.7	205	0.7	47	0.2	4	0.4	–	–	–	–	–	–
Liby.Arab Jam.	11	0.1	20	0.1	128	0.4	50	0.2	0	0.0	5	0.3	8	0.3	13	0.2
Morocco	18	0.1	12	0.0	11	0.0	5	0.0	–	–	–	–	0	0.0	0	0.0
Tunisia	2	0.0	11	0.0	12	0.0	17	0.1	–	–	–	–	–	–	–	–
Above listed	106	0.6	279	0.9	356	1.2	118	0.4	4	0.4	5	0.3	8	0.3	13	0.2
South Africa	223	1.3	377	1.2	56	0.2	83	0.3	–	–	16	1.1	1.	0.0	–	–
Other Africa	463	2.7	697	2.2	641	2.1	730	2.6	31	3.1	8	0.5	49	1.7	66	1.0
Far East	5252	30.1	10829	33.7	14055	45.5	12000	41.9	64	6.4	24	1.6	99	3.5	1694	25.0
China	1536	8.8	2852	8.9	5530	17.9	2929	10.2	.	.	.	.	.	.	.	.
India	131	0.7	739	2.3	297	1.0	833	2.9	.	.	.	.	.	.	.	.
Korea,Rep.of	456	2.6	1877	5.8	2287	7.4	1071	3.7	.	.	.	.	.	.	.	.
Above listed	2123	12.2	5469	17.0	8113	26.3	4833	16.9	.	.	.	.	.	.	.	.
Japan	.	.	.	.	.	.	.	.	64	6.4	24	1.6	99	3.5	1694	25.0
Other Far East	3129	17.9	5360	16.7	5941	19.2	7168	25.1	.	.	.	.	.	.	.	.
Middle East	608	3.5	3223	10.0	3733	12.1	5810	20.3	277	27.9	149	9.9	501	17.4	1036	15.3
Egypt	35	0.2	26	0.1	73	0.2	101	0.4	12	1.2	0	0.0	28	1.0	16	0.2
Saudi Arabia	51	0.3	436	1.4	1063	3.4	3133	11.0	56	5.6	15	1.0	131	4.6	808	11.9
Above listed	86	0.5	461	1.4	1136	3.7	3234	11.3	68	6.9	15	1.0	159	5.5	825	12.2
Other Middle East	522	3.0	2761	8.6	2597	8.4	2576	9.0	209	21.0	134	8.9	342	11.9	212	3.1
Oceania	594	3.4	1248	3.9	618	2.0	940	3.3	8	0.8	70	4.6	76	2.7	103	1.5
Australia	376	2.2	727	2.3	334	1.1	573	2.0	5	0.5	50	3.3	74	2.6	92	1.4
New Zealand	204	1.2	495	1.5	270	0.9	351	1.2	2	0.2	16	1.0	2.	0.1	11	0.2
Other Oceania	14	0.1	26	0.1	14	0.0	15	0.1	0	0.0	5	0.3	0	0.0	0	0.0
North America	5907	33.8	6675	20.8	5996	19.4	4170	14.6	87	8.8	858	57.0	1058	36.8	924	13.6
Canada	344	2.0	912	2.8	409	1.3	259	0.9	0	0.0	4	0.3	12	0.4	45	0.7
United States	5563	31.8	5762	17.9	5587	18.1	3911	13.7	87	8.8	854	56.7	1046	36.4	879	13.0
N. America unalloc	–	–	–	–	0	0.0	0	0.0	0	0.0	0	0.0	–	–	0	0.0
Other America	1566	9.0	4350	13.6	2415	7.8	1478	5.2	2	0.2	118	7.9	7	0.2	102	1.5
Argentina	422	2.4	1050	3.3	163	0.5	130	0.5	–	–	6	0.4	4	0.1	1	0.0
Brazil	191	1.1	1547	4.8	214	0.7	41	0.1	–	–	64	4.2	0	0.0	0	0.0
Mexico	22	0.1	119	0.4	457	1.5	271	0.9	–	–	5	0.3	–	–	–	–
Venezuela	313	1.8	593	1.8	627	2.0	426	1.5	1	0.1	33	2.2	1	0.0	17	0.2
Above listed	948	5.4	3309	10.3	1460	4.7	868	3.0	1	0.1	108	7.1	5	0.2	18	0.3
Oth. Oth. America	618	3.5	1041	3.2	954	3.1	610	2.1	1	0.1	11	0.7	2	0.1	84	1.2
Western Europe	2286	13.1	2844	8.9	1342	4.3	855	3.0	36	3.6	59	3.9	97	3.4	759	11.2
B.L.E.U.	211	1.2	220	0.7	124	0.4	48	0.2	0	0.1	13	0.8	39	1.3	85	1.2
Denmark	16	0.1	38	0.1	10	0.0	8	0.0	1	0.1	0	0.0	0	0.0	0	0.0
France	7	0.0	17	0.1	22	0.1	18	0.1	–	–	0	0.0	–	–	0	0.0
Germany,Fed.Rep.	49	0.3	227	0.7	193	0.6	74	0.3	5	0.5	2	0.1	2	0.1	5	0.1
Ireland	6	0.0	6	0.0	9	0.0	1	0.0	–	–	2	0.1	3	0.1	0	0.0
Italy	629	3.6	221	0.7	52	0.2	20	0.1	–	–	10	0.7	0	0.0	6	0.1
Netherlands	50	0.3	35	0.1	57	0.2	11	0.0	2	0.2	3	0.2	1	0.0	6	0.1
United Kingdom	70	0.4	290	0.9	156	0.5	81	0.3	15	1.5	9	0.6	40	1.4	50	0.7
EEC(9)	1038	5.9	1054	3.3	623	2.0	260	0.9	23	2.3	38	2.5	86	3.0	145	2.1
Finland	13	0.1	11	0.0	3	0.0	7	0.0	–	–	–	–	–	–	0	0.0
Iceland	8	0.0	–	–	0	0.0	–	–	0	0.0	–	–	–	–	–	–
Norway	95	0.5	243	0.8	33	0.1	223	0.8	–	–	–	–	2	0.1	589	8.7
Sweden	85	0.5	141	0.4	83	0.3	32	0.1	–	–	0	0.0	5	0.2	0	0.0
Northern Europe	201	1.2	394	1.2	118	0.4	262	0.9	0	0.0	0	0.0	7	0.2	589	8.7
Greece	251	1.4	506	1.6	162	0.5	90	0.3	0	0.0	0	0.0	0	0.0	8	0.1
Portugal	65	0.4	215	0.7	135	0.4	50	0.2	–	–	–	–	–	–	9	0.1
Spain	456	2.6	85	0.3	36	0.1	21	0.1	5	0.5	–	–	0	0.0	6	0.1
Turkey	37	0.2	318	1.0	164	0.5	145	0.5	0	0.0	20	1.3	–	–	1	0.0
Yugoslavia	218	1.2	208	0.6	19	0.1	7	0.0	6	0.6	0	0.0	1	0.0	0	0.0
Southern Europe	1027	5.9	1333	4.2	516	1.7	312	1.1	11	1.1	20	1.4	1	0.1	24	0.3
Austria	1	0.0	7	0.0	6	0.0	4	0.0	–	–	0	0.0	–	–	2	0.0
Switzerland	10	0.1	56	0.2	79	0.3	15	0.1	–	–	0	0.0	–	–	–	–
Central Europe	11	0.1	63	0.2	85	0.3	19	0.1	–	–	0	0.0	1	0.0	2	0.0
W.Europe unalloc	9	0.1	1	0.0	1	0.0	2	0.0	1	0.1	0	0.0	2	0.1	0	0.0
Eastern Europe	466	2.7	1576	4.9	1664	5.4	2421	8.5	139	14.0	2	0.1	87	3.0	15	0.2
Albania	4	0.0	2	0.0	3	0.0	6	0.0	–	–	–	–	–	–	–	–
Bulgaria	20	0.1	22	0.1	9	0.0	11	0.0	1	0.1	–	–	–	–	–	–
Czechoslovakia	–	–	1	0.0	2	0.0	1	0.0	1	0.1	1	0.1	–	–	–	–
German Dem.Rep.	84	0.5	60	0.2	12	0.0	11	0.0	–	–	–	–	–	–	–	–
Hungary	4	0.0	2	0.0	1	0.0	6	0.0	0	0.0	–	–	0	0.0	–	–
Poland	16	0.1	90	0.3	30	0.1	3	0.0	–	–	1	0.1	0	0.0	–	–
Romania	119	0.7	82	0.3	91	0.3	5	0.0	–	–	–	–	–	–	–	–
Above listed	248	1.4	259	0.8	148	0.5	41	0.1	2	0.2	2	0.1	0	0.0	–	–
USSR	218	1.2	1317	4.1	1516	4.9	2380	8.3	137	13.8	0	0.0	86	3.0	15	0.2
E.Europe unalloc	–	–	–	–	–	–	–	–	–	–	–	–	–	–	–	–
Unallocated	–	–	5	0.0	–	–	–	–	345	34.8	197	13.1	894	31.1	2069	30.5
World	17471	100.0	32103	100.0	30876	100.0	28606	100.0	992	100.0	1507	100.0	2877	100.0	6783	100.0

Another important destination region for Japanese exports was the Middle East. Between 1970 and 1982, Japan expanded its exports to this region from 0.61 million to 5.81 million tonnes or by a factor of more than nine, and their share in total Japanese exports of steel products increased from 3.5 to 20.3 per cent.

While 33.8 per cent of Japanese steel exports was directed to North America in 1970, this destination only represented 14.6 per cent of the total steel exported by Japan in 1982.

The share of Japanese steel exports to the USSR grew significantly, attaining 8.2 per cent in 1982 compared with 1.2 per cent in 1970. Concurrently, the volume of exports increased from 0.22 million to 2.38 million tonnes.

Japan exported a smaller proportion of its exports to the other regions and, as a rule during the period under review, the volume of its exports to those regions decreased in terms of both volume and percentage of total Japanese steel exports.

Changes in the geographic orientation of exports from the Far East (excluding Japan)

The steel exports of the Far East grew dynamically from 1970 (0.99 million tonnes) to 1982 (6.78 million tonnes), the main part of those exports being directed to Japan. Between 1970 and 1982 its exports of steel products to Japan increased from 0.06 million to 1.69 million tonnes. During the same period, the share of the region's steel exports to Japan grew from 6.4 to 25 per cent (table 2.6).

A large proportion of Far Eastern exports went to the Middle East, but although the volume of these exports increased from 0.28 million to 1.04 million tonnes, their share in total Far East exports declined from 27.9 to 15.3 per cent between 1970 and 1982.

The Far East increased the volume of its exports as well as their shares in total exports to the following regions during the period under review:

- Oceania: Exports - from 0.01 million to 0.1 million tonnes.
  Share - from 0.8 to 1.5 per cent;

- North America: Exports - from 0.09 million to 0.92 million tonnes.
  Share - from 8.8 to 13.6 per cent;

- Latin America: Exports - from practically zero to 0.10 million tonnes. In 1982 its share accounted for 1.5 per cent; and

- Northern Europe: Exports - from zero to 0.59 million tonnes.
  Share - from zero to 8.7 per cent.

During the period 1970-1982, there was a decrease in Far Eastern steel exports to the following regions: Africa (from 3.5 to 1.2 per cent), EEC (from 2.3 to 2.1 per cent), southern Europe (from 1.1 to 0.3 per cent), eastern Europe (from 0.2 per cent to zero) and the USSR (from 13.8 to 0.2 per cent).

In the absence of information concerning the destination of a part of the steel products exported from the Far East, in particular from the Republic of Korea these exports were considered unallocated. In 1982, 30.5 per cent of Far Eastern steel exports were unallocated.

Changes in the geographic orientation of exports from the countries of northern Europe

Between 1970 and 1973, the steel exports of the north European countries increased from 1.84 million to 2.33 million tonnes. From 1974 to 1978, the volume of their exports fluctuated between 2 million and 3.28 million tonnes and in 1979 there was a peak, when they attained 3.38 million tonnes (table 2.1). During the following three years their exports of steel products remained at around 3 million tonnes.

While the major part of the north European countries' exports continued to be directed to EEC, the share of these exports decreased from 77.3 to 64.9 per cent (table 2.5). This was due in particular to the extension of exports of steel products to North America and to the Middle East. The shares in total steel exports of the countries of northern Europe in those two regions grew from 4.7 to 13.6 per cent and from 0.7 to 4.2 per cent, respectively.

The share of exports directed to Africa, the Far East and southern Europe also increased slightly, from 0.8 to 1.7 per cent, from 3.0 to 3.2 per cent and from 4.1 to 5.4 per cent, respectively, during the period.

The share of exports directed to Oceania, Latin America, central Europe and eastern Europe declined.

Changes in the geographic orientation of exports from the countries of southern Europe

During the period, the south European countries dynamically increased their exports, from 0.62 to 6.13 million tonnes, with only one interruption, between 1974 and 1975, when the volume of exports declined from 2.48 million tonnes in 1973 to 1.55 million tonnes in 1974.

The main feature of the changes in the geographic orientation of their exports was that EEC lost its first place as the major destination of steel exports from southern Europe. The EEC share in the total steel exports of this region decreased from 41.1 per cent in 1970 to 18.9 per cent in 1982 (table 2.7).

The south European countries recorded a considerable increase in their exports of steel products to the Middle East, with a volume growth of from 0.04 million to 1.95 million tonnes and a share increase of from 5.7 to 31.8 per cent between 1970 and 1982.

Exports of this region to Africa also increased significantly, from 0.05 million tonnes in 1970 to 1.21 million tonnes in 1982. Thus Africa moved into third place as a destination of steel exports from south European countries, its share growing from 7.4 to 19.8 per cent.

Table 2.7. Changes in the geographic orientation of exports of steel products

Years / Importers	South European countries								Central European countries							
	1970		1974		1978		1982		1970		1974		1978		1982	
	1000t	%	1000t	%	1000t	%	1000t	%	1000t	%	1000t	%	1000t	%	1000t	%
Africa	46	7.4	106	6.9	498	10.5	1214	19.8	9	0.7	15	0.9	39	1.6	33	1.2
Algeria	4	0.6	23	1.5	149	3.2	547	8.9	1	0.0	8	0.5	26	1.0	13	0.4
Liby.Arab Jam.	2	0.3	40	2.6	30	0.6	27	0.4	0	0.0	1	0.0	0	0.0	1	0.0
Morocco	9	1.4	18	1.2	190	4.0	360	5.9	1	0.1	1	0.1	0	0.0	1	0.0
Tunisia	0	0.0	0	0.0	24	0.5	122	2.0	2	0.2	0	0.0	6	0.3	11	0.4
Above listed	14	2.3	81	5.2	393	8.3	1056	17.2	3	0.3	10	0.6	33	1.3	25	0.9
South Africa	1	0.1	10	0.7	2	0.1	2	0.0	2	0.1	2	0.1	1	0.0	2	0.1
Other Africa	31	5.0	15	1.0	103	2.2	156	2.6	4	0.3	3	0.2	5	0.2	6	0.2
Far East	24	3.8	50	3.3	410	8.7	201	3.3	10	0.8	11	0.6	24	1.0	12	0.4
China	22	3.5	39	2.5	182	3.8	38	0.6	5	0.4	2	0.1	12	0.5	0	0.0
India	1	0.1	4	0.2	2	0.1	105	1.7	3	0.1	4	0.2	5	0.2	8	0.3
Korea,Rep.of	-	-	0	0.0	21	0.4	1	0.0	-	-	0	0.0	0	0.0	0	0.0
Above listed	23	3.7	43	2.8	205	4.3	144	2.3	8	0.6	5	0.3	17	0.7	9	0.3
Japan	-	-	0	0.0	34	0.7	0	0.0	0	0.0	1	0.0	0	0.0	0	0.0
Other Far East	1	0.2	7	0.5	171	3.6	57	0.9	2	0.2	5	0.3	7	0.3	3	0.1
Middle East	36	5.7	175	11.3	673	14.2	1947	31.8	31	2.4	72	4.1	31	1.3	120	4.2
Egypt	12	1.9	34	2.2	46	1.0	159	2.6	1	0.1	4	0.3	7	0.3	32	1.1
Saudi Arabia	0	0.0	3	0.2	53	1.1	117	1.9	0	0.0	0	0.0	1	0.1	3	0.1
Above listed	12	1.9	37	2.4	99	2.1	276	4.5	1	0.1	5	0.3	8	0.3	35	1.2
Other Middle East	24	3.8	137	8.9	574	12.1	1671	27.3	29	2.3	67	3.8	23	0.9	85	2.9
Oceania	1	0.1	0	0.0	3	0.1	1	0.0	3	0.2	4	0.2	2	0.1	3	0.1
Australia	1	0.1	0	0.0	3	0.1	1	0.0	3	0.2	3	0.2	1	0.1	3	0.1
New Zealand	-	-	0	0.0	0	0.0	0	0.0	0	0.0	0	0.0	0	0.0	0	0.0
Other Oceania	0	0.0	0	0.0	-	-	0	0.0	0	0.0	0	0.0	-	-	0	0.0
North America	63	10.1	145	9.4	656	13.9	524	8.5	31	2.4	35	2.0	24	1.0	74	2.6
Canada	3	0.4	2	0.1	15	0.3	26	0.4	11	0.9	9	0.5	7	0.3	1	0.0
United States	60	9.7	143	9.2	641	13.5	497	8.1	19	1.5	26	1.5	17	0.7	73	2.5
N. America unalloc.	0	0.0	0	0.0	-	-	0	0.0	0	0.0	0	0.0	0	0.0	-	-
Other America	20	3.3	82	5.3	241	5.1	227	3.7	20	1.5	19	1.1	22	0.9	33	1.1
Argentina	6	1.0	2	0.2	11	0.2	16	0.3	5	0.4	5	0.3	10	0.4	2	0.1
Brazil	1	0.1	52	3.4	61	1.3	61	1.0	1	0.1	6	0.3	6	0.2	8	0.3
Mexico	0	0.0	2	0.1	19	0.4	50	0.8	1	0.1	1	0.0	0	0.0	1	0.0
Venezuela	0	0.0	14	0.9	103	2.2	16	0.3	1	0.1	1	0.0	0	0.0	0	0.0
Above listed	7	1.2	71	4.6	194	4.1	143	2.3	9	0.7	12	0.7	16	0.7	11	0.4
Oth. Oth. America	13	2.1	12	0.7	46	1.0	85	1.4	11	0.9	6	0.4	6	0.2	21	0.7
Western Europe	279	45.0	479	31.0	1534	32.4	1513	24.7	964	74.0	1244	71.2	1916	77.8	2010	69.6
B.L.E.U.	4	0.6	49	3.1	70	1.5	91	1.5	5	0.4	17	1.0	27	1.1	28	1.0
Denmark	0	0.0	4	0.2	10	0.2	7	0.1	15	1.2	32	1.8	53	2.2	48	1.6
France	74	12.0	75	4.9	272	5.7	399	6.5	22	1.7	39	2.2	135	5.5	141	4.9
Germany,Fed.Rep.	76	12.3	155	10.1	453	9.6	365	6.0	606	46.5	560	32.0	984	40.0	1049	36.3
Ireland	0	0.0	2	0.1	25	0.5	6	0.1	1	0.1	1	0.1	3	0.1	6	0.2
Italy	75	12.0	94	6.1	199	4.2	118	1.9	110	8.4	210	12.0	247	10.0	317	11.0
Netherlands	12	1.9	6	0.4	36	0.8	36	0.6	20	1.5	47	2.7	66	2.7	79	2.7
United Kingdom	13	2.2	49	3.2	112	2.4	134	2.2	66	5.1	100	5.7	111	4.5	96	3.3
EEC(9)	255	41.1	434	28.1	1177	24.9	1156	18.9	844	64.8	1005	57.5	1626	66.0	1764	61.0
Finland	7	1.1	3	0.2	18	0.4	38	0.6	16	1.3	22	1.3	10	0.4	14	0.5
Iceland	0	0.0	0	0.0	-	-	1	0.0	0	0.0	0	0.0	0	0.0	0	0.0
Norway	1	0.1	4	0.3	32	0.7	63	1.0	9	0.7	10	0.6	19	0.8	17	0.6
Sweden	1	0.1	5	0.4	131	2.8	93	1.5	39	3.0	79	4.5	76	3.1	77	2.7
Northern Europe	9	1.4	13	0.8	181	3.8	195	3.2	65	5.0	112	6.4	105	4.3	108	3.7
Greece	.	.	.	.	.	.	.	.	13	1.0	20	1.2	8	0.3	9	0.3
Portugal	.	.	.	.	.	.	.	.	8	0.6	5	0.3	8	0.3	4	0.1
Spain	.	.	.	.	.	.	.	.	5	0.4	11	0.6	3	0.1	9	0.3
Turkey	.	.	.	.	.	.	.	.	1	0.1	29	1.6	8	0.3	5	0.2
Yugoslavia	.	.	.	.	.	.	.	.	26	2.0	62	3.6	158	6.4	111	3.8
Southern Europe	.	.	.	.	.	.	.	.	55	4.2	127	7.3	185	7.5	138	4.8
Austria	2	0.3	7	0.4	21	0.4	19	0.3	.	.	.	.	.	.	.	.
Switzerland	9	1.4	14	0.9	132	2.8	132	2.2	.	.	.	.	.	.	.	.
Central Europe	11	1.7	21	1.4	153	3.2	151	2.5	.	.	.	.	.	.	.	.
W.Europe unalloc	5	0.8	11	0.7	23	0.5	10	0.2	1	0.0	1	0.1	0	0.0	0	0.0
Eastern Europe	152	24.5	508	32.9	672	14.2	500	8.2	236	18.1	350	20.0	405	16.4	604	20.9
Albania	3	0.5	18	1.2	15	0.3	29	0.5	0	0.0	0	0.0	0	0.0	1	0.0
Bulgaria	48	7.7	38	2.4	54	1.1	47	0.8	29	2.2	12	0.7	11	0.4	28	1.0
Czechoslovakia	1	0.2	7	0.5	11	0.2	14	0.2	22	1.7	16	0.9	11	0.4	7	0.3
German Dem.Rep.	10	1.6	28	1.8	14	0.3	38	0.6	13	1.0	18	1.1	21	0.8	15	0.5
Hungary	13	2.1	10	0.7	20	0.4	11	0.2	37	2.8	65	3.7	22	0.9	12	0.4
Poland	3	0.5	113	7.3	44	0.9	42	0.7	9	0.7	61	3.5	50	2.0	4	0.1
Romania	42	6.1	66	4.3	77	1.6	46	0.8	27	2.1	23	1.3	29	1.2	7	0.2
Above listed	120	19.3	281	18.2	235	5.0	228	3.7	138	10.6	197	11.2	143	5.8	74	2.6
USSR	32	5.2	228	14.7	437	9.2	272	4.4	98	7.5	153	8.8	262	10.6	529	18.3
E.Europe unalloc	0	0.0	0	0.0	-	-	-	-	0	0.0	0	0.0	-	-	-	-
Unallocated	0	0.0	0	0.0	45	0.9	1	0.0	0	0.0	0	0.0	0	0.0	-	-
World	621	100.0	1547	100.0	4732	100.0	6127	100.0	1304	100.0	1749	100.0	2463	100.0	2889	100.0

The south European countries increased the volume of their steel exports to the Far East (from 0.02 million to 0.2 million tonnes), to North America (from 0.06 million to 0.52 million tonnes), to eastern Europe (from 0.12 million to 0.23 million tonnes) and to the USSR (from 0.03 million to 0.27 million tonnes). Nevertheless, the region's share in total steel exports declined from 3.8 to 3.3 per cent, from 10.1 to 8.5 per cent, from 19.3 to 10.6 per cent and from 5.2 to 4.4 per cent, respectively.

The countries of southern Europe increased the volume of their exports as well as their share in total exports to the following regions:

- Latin America: Exports - from 0.02 million to 0.23 million tonnes. Share - from 3.3 to 3.7 per cent;

- Northern Europe: Exports - from 0.01 million to 0.20 million tonnes. Share - from 1.4 to 3.2 per cent; and

- Central Europe: Exports - from 0.01 million to 0.15 million tonnes. Share - from 1.7 to 2.5 per cent.

Changes in the geographic orientation of exports from the countries of central Europe

The volume of steel exports from the central European countries grew from 1.30 million tonnes in 1970 to 2.89 million tonnes in 1982. Next to the growing trend appearing in 1973, 1977 and 1980, some small decreases occurred compared with previous years. The volume of exports reached its highest point in 1981 (3.10 million tonnes) (table 2.1).

EEC was and remains the main recipient of the steel exports of central Europe (table 2.7). In 1970, 64.8 per cent of those exports went to EEC. During the period under review this share slightly declined, reaching 61.0 per cent in 1982. The highest share was recorded in 1981 (66.0 per cent). Meanwhile, the volume of exports increased from 0.84 million to 1.76 million tonnes between 1970 and 1982.

Another considerable portion of the central European exports was directed to the USSR. The share of the region's exports to the USSR grew from 7.5 per cent in 1970 to 18.3 per cent in 1982.

The central European countries increased the share in their total exports of exports to Africa (from 0.7 to 1.2 per cent), to the Middle East (from 2.4 to 4.2 per cent), to North America (from 2.4 to 2.6 per cent) and to southern Europe (from 4.2 to 4.8 per cent).

The other trend brought out by the table is the decline of the share of exports to the Far East (from 0.8 to 0.4 per cent), to Oceania (from 0.2 to 0.1 per cent), to Latin America (from 1.5 to 1.1 per cent), to northern Europe (from 5.0 to 3.7 per cent) and to eastern Europe (from 10.5 to 2.6 per cent).

The total volume of steel products exported by north European, south European and central European countries together, as well as the respective shares in total steel exports to various countries and regions, are reflected in table 2.8.

Table 2.8. Changes in the geographic orientation of exports of steel products

| Years | Other west European countries | | | | | | | |
| Importers | 1970 | | 1974 | | 1978 | | 1982 | |
	1000t	%	1000t	%	1000t	%	1000t	%
Africa	70	1.5	169	2.5	615	5.3	1298	9.9
Algeria	4	0.1	39	0.6	241	2.1	576	4.4
Liby.Arab Jam.	2	0.0	47	0.7	31	0.3	41	0.3
Morocoo	14	0.3	20	0.3	191	1.7	361	2.7
Tunisia	2	0.0	1	0.0	30	0.3	138	1.0
Above listed	22	0.5	107	1.5	493	4.3	1117	8.5
South Africa	7	0.2	26	0.4	5	0.0	9	0.1
Other Africa	40	0.9	37	0.5	118	1.0	173	1.3
Far East	94	2.0	102	1.5	608	5.3	316	2.4
China	64	1.4	53	0.8	228	2.0	48	0.4
India	8	0.2	19	0.3	18	0.2	157	1.2
Korea,Rep.of	0	0.0	0	0.0	22	0.2	3	0.0
Above listed	72	1.6	72	1.0	269	2.3	208	1.6
Japan	5	0.1	5	0.1	61	0.5	3	0.0
Other Far East	17	0.4	25	0.4	279	2.4	104	0.8
Middle East	79	1.7	283	4.1	824	7.1	2197	16.7
Egypt	13	0.3	40	0.6	55	0.5	192	1.5
Saudi Arabia	0	0.0	4	0.1	61	0.5	184	1.4
Above listed	13	0.3	44	0.6	116	1.0	376	2.9
Other Middle East	65	1.4	238	3.4	708	6.1	1820	13.8
Oceania	13	0.3	16	0.2	12	0.1	11	0.1
Australia	13	0.3	15	0.2	12	0.1	10	0.1
New Zealand	0	0.0	1	0.0	0	0.0	1	0.0
Other Oceania	0	0.0	0	0.0	–		–	0.0
North America	180	3.9	338	4.9	1113	9.6	1016	7.7
Canada	29	0.6	36	0.5	44	0.4	48	0.4
United States	152	3.3	302	4.4	1068	9.2	967	7.3
N. America unalloc	0	0.0	0	0.0	0	0.0	0	0.0
Other America	70	1.5	175	2.5	352	3.0	293	2.2
Argentina	19	0.4	16	0.2	24	0.2	33	0.3
Brazil	13	0.3	105	1.5	74	0.6	72	0.5
Mexico	6	0.1	12	0.2	28	0.2	59	0.4
Venezuela	4	0.1	17	0.2	106	0.9	18	0.1
Above listed	41	0.9	150	2.2	232	2.0	182	1.4
Oth. Oth. America	29	0.6	25	0.4	120	1.0	111	0.8
Western Europe	3593	78.5	4772	69.1	6717	58.1	6830	51.8
B.L.E.U.	48	1.1	137	2.0	206	1.8	182	1.4
Denmark	359	7.8	366	5.3	338	2.9	429	3.3
France	215	4.7	317	4.6	614	5.3	723	5.5
Germany,Fed.Rep.	1155	25.2	1130	16.4	2076	18.0	2059	15.6
Ireland	4	0.1	8	0.1	38	0.3	27	0.2
Italy	262	5.7	360	5.2	487	4.2	509	3.9
Netherlands	146	3.2	232	3.4	354	3.1	350	2.7
United Kingdom	331	7.2	457	6.6	775	6.7	635	4.8
EEC(9)	2520	55.1	3006	43.5	4888	42.3	4914	37.3
Finland	162	3.6	225	3.3	162	1.4	196	1.5
Iceland	6	0.1	14	0.2	17	0.1	20	0.1
Norway	191	4.2	247	3.6	289	2.5	348	2.6
Sweden	216	4.7	421	6.1	554	4.8	427	3.2
Northern Europe	576	12.6	907	13.1	1021	8.8	992	7.5
Greece	35	0.8	48	0.7	68	0.6	40	0.3
Portugal	55	1.2	49	0.7	69	0.6	137	1.0
Spain	37	0.8	208	3.0	27	0.2	91	0.7
Turkey	5	0.1	108	1.6	49	0.4	75	0.6
Yugoslavia	140	3.1	166	2.4	199	1.7	208	1.6
Southern Europe	271	5.9	579	8.4	414	3.6	552	4.2
Austria	32	0.7	81	1.2	83	0.7	87	0.7
Switzerland	188	4.1	186	2.7	287	2.5	274	2.1
Central Europe	220	4.8	267	3.9	370	3.2	361	2.7
W.Europe unalloc	6	0.1	13	0.2	25	0.2	12	0.1
Eastern Europe	476	10.4	1051	15.2	1270	11.0	1220	9.3
Albania	3	0.1	18	0.3	16	0.1	31	0.2
Bulgaria	77	1.7	50	0.7	70	0.6	84	0.6
Czechoslovakia	25	0.6	26	0.4	24	0.2	23	0.2
German Dem.Rep.	49	1.1	74	1.1	53	0.5	61	0.5
Hungary	52	1.1	78	1.1	54	0.5	33	0.3
Poland	23	0.5	288	4.2	134	1.2	57	0.4
Romania	72	1.6	101	1.5	115	1.0	55	0.4
Above listed	302	6.6	636	9.2	465	4.0	343	2.6
USSR	173	3.8	415	6.0	805	7.0	876	6.6
E.Europe unalloc	0	0.0	0	0.0	–		–	–
Unallocated	0	0.0	1	0.0	45	0.4	1	0.0
World	4575	100.0	6907	100.0	11556	100.0	13181	100.0

Changes in the geographic orientation of exports from the countries of
North America

The exports of the countries of North America showed a sharp decline over
the whole of the period 1970-1982, decreasing from 6.12 million to
2.94 million tonnes. As can be seen from table 2.1, there were, however, some
increases in exports during the period, for instance in 1974 and 1980, when
exports amounted to 4.25 million tonnes and 4.68 million tonnes, respectively.

Table 2.9 gives a picture of the geographic orientation of steel-product
exports from the North American countries. The figures shown are evidence of
a decline in the volume of their exports to almost all areas of the world with
the exception of the Middle East. Exports to the Middle East increased from
0.07 million to 0.63 million tonnes between 1970 and 1982, and their share in
total exports to the Middle East grew from 1.2 to 21.4 per cent.

The North American countries also increased the share in their total
exports of steel products of exports to the following regions:

- Africa: from 3.3 to 5.0 per cent;
- Far East: from 13.5 to 23.3 per cent;
- Japan: from 0.3 to 0.5 per cent; and
- Latin America: from 18.9 to 25.7 per cent.

There was a remarkable decrease in volume as well as in the share of
exports to EEC. The volume of exports decreased from 2.96 million to
0.43 million tonnes and their share declined from 48.3 to 14.7 per cent
between 1970 and 1982.

A decline was recorded by the countries of North America in their exports
and in the share in their total exports of exports to the following regions:

- Oceania: Exports - from 0.06 million to 0.02 million tonnes.
  Share - from 0.9 to 0.5 per cent;

- Northern Europe: Exports - from 0.09 million to 0.02 million tonnes.
  Share - from 1.5 to 0.5 per cent;

- Southern Europe: Exports - from 0.58 million to 0.23 million tonnes.
  Share - from 9.5 to 7.7 per cent; and

- Central Europe: Exports - from 0.08 million to 0.02 million tonnes.
  Share - from 1.2 to 0.7 per cent.

The North American countries had been directing only a small proportion
of their steel exports to eastern Europe and the USSR and, by 1982, they had
practically ceased all steel exports to those countries.

Changes in the geographic orientation of exports from the countries of
Latin America

While steel exports from Latin American countries grew from 0.54 million
to 3.13 million tonnes during the period from 1970 to 1977, they were
characterized by significant fluctuations. Between 1970 and 1972 there was an

Table 2.9. Changes in the geographic orientation of exports of steel products

Years / Importers	North America 1970 1000t	%	1974 1000t	%	1978 1000t	%	1982 1000t	%	Latin America 1970 1000t	%	1974 1000t	%	1978 1000t	%	1982 1000t	%
Africa	204	3.3	293	6.9	101	4.6	148	5.0	130	23.9	14	4.8	94	6.6	195	6.2
Algeria	32	0.5	50	1.2	12	0.5	36	1.2	117	21.5	4	1.5	35	2.5	84	2.7
Liby.Arab Jam.	12	0.2	8	0.2	4	0.2	2	0.1	-	-	-	-	2	0.1	1	0.0
Morocco	14	0.2	1	0.0	0	0.0	0	0.0	4	0.8	-	-	3	0.2	0	0.0
Tunisia	3	0.1	2	0.0	5	0.2	1	0.0	9	1.6	-	-	0	0.0	0	0.0
Above listed	61	1.0	61	1.4	20	0.9	39	1.3	130	23.9	4	1.5	40	2.8	85	2.7
South Africa	57	0.9	104	2.4	5	0.2	3	0.1	0	0.0	1	0.2	0	0.0	1	0.0
Other Africa	86	1.4	128	3.0	75	3.5	106	3.6	0	0.0	9	3.0	54	3.8	109	3.5
Far East	836	13.7	605	14.2	555	25.4	699	23.8	0	0.0	0	0.2	204	14.3	627	20.0
China	0	0.0	2	0.0	0	0.0	79	2.7	-	-	-	-	30	2.1	36	1.1
India	213	3.5	14	0.3	50	2.3	136	4.6	-	-	-	-	-	-	81	2.6
Korea,Rep.of	16	0.3	18	0.4	104	4.8	28	1.0	-	-	-	-	95	6.7	2	0.1
Above listed	229	3.7	33	0.8	155	7.1	243	8.3	-	-	-	-	125	8.8	118	3.8
Japan	16	0.3	12	0.3	54	2.4	14	0.5	0	0.0	0	0.0	26	1.8	221	7.1
Other Far East	592	9.7	559	13.2	346	15.8	441	15.0	-	-	0	0.1	53	3.7	288	9.2
Middle East	72	1.2	280	6.6	199	9.1	629	21.4	8	1.5	5	1.8	50	3.5	529	16.9
Egypt	4	0.1	24	0.6	48	2.2	78	2.6	-	-	-	-	-	-	34	1.1
Saudi Arabia	3	0.0	25	0.6	75	3.4	153	5.2	0	0.0	-	-	15	1.0	213	6.8
Above listed	7	0.1	49	1.2	123	5.6	231	7.9	0	0.0	-	-	15	1.0	247	7.9
Other Middle East	66	1.1	230	5.4	76	3.5	398	13.5	8	1.5	5	1.8	36	2.5	282	9.0
Oceania	55	0.9	67	1.6	17	0.8	15	0.5	0	0.0	0	0.0	1	0.1	22	0.7
Australia	27	0.4	49	1.2	9	0.4	9	0.3	0	0.0	0	0.0	1	0.1	22	0.7
New Zealand	27	0.4	16	0.4	8	0.4	6	0.2	-	-	0	0.0	-	-	0	0.0
Other Oceania	1	0.0	2	0.0	1	0.0	1	0.0	-	-	-	-	-	-	0	0.0
North America	.	.	.	.	.	.	.	.	277	50.8	257	86.1	736	51.6	1021	32.7
Canada	.	.	.	.	.	.	.	.	0	0.0	2	0.7	23	1.6	59	1.9
United States	.	.	.	.	.	.	.	.	277	50.8	255	85.4	713	50.0	962	30.8
N. America unalloc.	.	.	.	.	.	.	.	.	-	-	0	0.0	0	0.0	0	0.0
Other America	1157	18.9	2172	51.1	956	43.7	757	25.7	.	.	.	.	.	.	.	.
Argentina	457	7.5	142	3.3	9	0.4	16	0.6	.	.	.	.	.	.	.	.
Brazil	79	1.3	878	20.7	48	2.2	20	0.7	.	.	.	.	.	.	.	.
Mexico	163	2.7	442	10.4	409	18.7	344	11.7	.	.	.	.	.	.	.	.
Venezuela	80	1.3	305	7.2	172	7.9	66	2.2	.	.	.	.	.	.	.	.
Above listed	779	12.7	1766	41.5	638	29.2	446	15.2	.	.	.	.	.	.	.	.
Oth. Oth. America	379	6.2	406	9.5	318	14.6	311	10.5	.	.	.	.	.	.	.	.
Western Europe	3707	60.6	770	18.1	318	14.6	693	23.5	129	23.7	21	6.9	324	22.7	716	22.9
B.L.E.U.	234	3.8	71	1.7	40	1.8	50	1.7	11	2.0	7	2.4	30	2.1	221	7.1
Denmark	5	0.1	4	0.1	1	0.1	0	0.0	-	-	-	-	-	-	26	0.8
France	479	7.8	24	0.6	19	0.9	16	0.6	0	0.0	1	0.2	0	0.0	1	0.0
Germany,Fed.Rep.	702	11.5	20	0.5	7	0.3	99	3.3	6	1.2	1	0.2	47	3.3	25	0.8
Ireland	5	0.1	2	0.0	1	0.1	1	0.0	-	-	-	-	-	-	21	0.7
Italy	590	9.6	200	4.7	95	4.4	157	5.3	44	8.1	3	0.9	1	0.1	187	6.0
Netherlands	12	0.2	35	0.8	14	0.6	4	0.1	13	2.3	0	0.0	51	3.6	39	1.2
United Kingdom	931	15.2	141	3.3	86	3.9	103	3.5	6	1.0	5	1.5	23	1.6	69	2.2
EEC(9)	2958	48.3	497	11.7	264	12.1	430	14.6	80	14.7	16	5.2	152	10.6	589	18.8
Finland	32	0.5	1	0.0	2	0.1	3	0.1	2	0.4	-	-	0	0.0	1	0.0
Iceland	1	0.0	0	0.0	0	0.0	0	0.0	-	-	-	-	-	-	-	-
Norway	20	0.3	12	0.3	2	0.1	1	0.0	2	0.4	-	-	-	-	15	0.5
Sweden	38	0.6	25	0.6	4	0.2	11	0.4	6	1.1	1	0.2	0	0.0	7	0.2
Northern Europe	90	1.5	39	0.9	7	0.3	16	0.5	10	1.8	1	0.2	0	0.0	22	0.7
Greece	207	3.4	55	1.3	20	0.9	70	2.4	2	0.4	0	0.0	156	10.9	14	0.4
Portugal	76	1.2	76	1.8	4	0.2	15	0.5	1	0.1	-	-	0	0.0	10	0.3
Spain	249	4.1	53	1.3	13	0.6	78	2.6	28	5.2	4	1.4	15	1.1	54	1.7
Turkey	52	0.8	40	0.9	1	0.0	62	2.1	9	1.6	-	-	-	-	-	-
Yugoslavia	0	0.0	4	0.1	1	0.1	1	0.0	-	-	-	-	-	-	-	-
Southern Europe	583	9.5	228	5.4	39	1.8	226	7.7	39	7.2	4	1.5	171	12.0	78	2.5
Austria	1	0.0	0	0.0	3	0.1	0	0.0	0	0.0	-	-	-	-	-	-
Switzerland	75	1.2	5	0.1	5	0.2	21	0.7	0	0.0	-	-	-	-	13	0.4
Central Europe	75	1.2	6	0.1	8	0.4	21	0.7	0	0.0	-	-	-	-	13	0.4
W.Europe unalloc	0	0.0	1	0.0	0	0.0	1	0.0	0	0.0	0	0.0	0	0.0	13	0.4
Eastern Europe	88	1.4	65	1.5	40	1.8	3	0.1	0	0.0	1	0.2	15	1.1	17	0.5
Albania	-	-	0	0.0	-	-	-	-	-	-	-	-	1	0.1	0	0.0
Bulgaria	-	-	0	0.0	0	0.0	1	0.0	-	-	-	-	1	0.1	0	0.0
Czechoslovakia	0	0.0	0	0.0	-	-	0	0.0	-	-	-	-	-	-	0	0.0
German Dem.Rep.	0	0.0	-	-	-	-	-	-	-	-	-	-	-	-	-	-
Hungary	-	-	-	-	-	-	0	0.0	-	-	-	-	-	-	-	-
Poland	0	0.0	24	0.6	20	0.9	-	-	-	-	-	-	-	-	-	-
Romania	39	0.6	17	0.4	12	0.5	-	-	0	0.0	0	0.0	1	0.1	-	-
Above listed	40	0.6	41	1.0	32	1.5	1	0.0	0	0.0	0	0.0	2	0.2	0	0.0
USSR	48	0.8	24	0.6	8	0.4	2	0.1	-	-	1	0.2	13	0.9	16	0.5
E.Europe unalloc	0	0.0	0	0.0	-	-	-	-	-	-	-	-	0	0.0	-	-
Unallocated	0	0.0	0	0.0	-	-	-	-	0	0.0	0	0.0	0	0.0	0	0.0
World	6120	100.0	4251	100.0	2186	100.0	2944	100.0	545	100.0	299	100.0	1424	100.0	3126	100.0

upward trend in the volume of exports, with an increase from 0.54 million to 0.78 million tonnes. This was followed by a sharp decrease in 1974 and 1975, with exports amounting to 0.3 million and 0.14 million tonnes, respectively. The year 1977 saw the beginning of a strong upward trend and by 1982 exports had increased from 0.5 million to 3.31 million tonnes.

Between 1970 and 1982, the major part of Latin American steel exports was directed to North America (see table 2.9). But the share decreased from 50.8 per cent in 1970 to 32.7 per cent in 1982.

In addition, the Latin American countries considerably increased their exports of steel products to the following regions:

- Far East: Exports - from zero to 0.41 million tonnes. Share - from zero to 12.9 per cent;

- Japan: Exports - from zero to 0.22 million tonnes. Share - from zero to 7.1 per cent.

- Middle East: Exports - from 0.01 million to 0.53 million tonnes. Share - from 1.5 to 16.9 per cent; and

- EEC: Exports - from 0.08 million to 0.59 million tonnes. Share - from 14.7 to 18.8 per cent.

While the countries of Latin America also began to export steel products to Oceania, central Europe and the USSR, it should be noted that the volume of those exports, as well as their share, were insignificant.

Between 1970 and 1982 the share of exports to Africa, northern Europe and southern Europe decreased from 23.9 to 6.2 per cent, from 1.8 to 0.7 per cent and from 7.2 to 2.5 per cent, respectively.

Changes in the geographic orientation of exports from the countries of eastern Europe

Total exports increased steadily from 1970 to 1973 and attained 7.79 million tonnes, compared with 6.13 million in 1970. Between 1974 and 1975, exports were steady, at around 7.5 million tonnes. In 1975 an upward trend started, with exports increasing from 7.49 million to 10.37 million tonnes by 1981. This was followed, in 1982, by a slight decrease, to 9.82 million tonnes (table 2.1).

Exports to other regions of the world went almost solely to western Europe, the USSR and the Middle East (table 2.10).

In the absence of complete data concerning the destination of steel exports from the German Democratic Republic and Romania, that part of those exports for which no destination is given is classified as unallocated. The share of unallocated exports decreased from 19.1 per cent in 1970 to 15.1 per cent in 1982. It may be assumed that most of those exports went to east European and west European countries.

Table 2.10. Changes in the geographic orientation of exports of steel products

	East European countries								USSR							
Years	1970		1974		1978		1982		1970		1974		1978		1982	
Importers	1000t	%	1000t	%	1000t	%	1000t	%	1000t	%	1000t	%	1000t	%	1000t	%
Africa	106	1.7	72	1.0	102	1.0	64	0.7	55	0.7	54	0.8	26	0.4	6	0.1
Algeria	6	0.1	16	0.2	17	0.2	8	0.1	15	0.2	31	0.5	–	–	–	–
Liby.Arab Jam.	22	0.4	8	0.1	18	0.2	–	–	23	0.3	17	0.2	–	–	–	–
Morocco	21	0.3	15	0.2	10	0.1	1	0.0	5	0.1	–	–	–	–	–	–
Tunisia	0	0.0	0	0.0	9	0.1	0	0.0	–	–	–	–	–	–	–	–
North Africa	48	0.8	39	0.5	54	0.6	9	0.1	43	0.6	47	0.7	–	–	–	–
South Africa	–	–	4	0.1	–	–	–	–	–	–	–	–	–	–	–	–
Other Africa	58	0.9	29	0.4	48	0.5	55	0.6	12	0.2	6	0.1	26	0.4	6	0.1
Far East	213	3.5	252	3.3	420	4.3	384	3.9	191	2.6	204	3.0	95	1.3	255	3.4
China	62	1.0	94	1.2	139	1.4	180	1.8	14	0.2	33	0.5	–	–	–	–
India	102	1.7	120	1.6	82	0.8	61	0.6	43	0.6	38	0.6	–	–	–	–
Korea,Rep.of	–	–	–	–	36	0.4	–	–	–	–	–	–	–	–	–	–
3 of Far East	164	2.7	214	2.8	257	2.6	241	2.5	57	0.8	71	1.0	–	–	–	–
Japan	16	0.3	1	0.0	56	0.6	–	–	–	–	–	–	–	–	–	–
Other Far East	33	0.5	37	0.5	106	1.1	143	1.5	134	1.8	133	1.9	95	1.3	255	3.4
Middle East	589	9.6	549	7.3	793	8.1	1128	11.5	435	5.9	238	3.5	68	0.9	75	1.0
Egypt	95	1.6	73	1.0	73	0.7	303	3.1	154	2.1	118	1.7	–	–	–	–
Saudi Arabia	7	0.1	2	0.0	4	0.0	38	0.4	12	0.2	2	0.0	–	–	–	–
2 of Middle East	102	1.7	75	1.0	77	0.8	342	3.5	167	2.2	120	1.8	–	–	–	–
Other Middle East	488	8.0	473	6.3	715	7.3	786	8.0	269	3.6	118	1.7	68	0.9	75	1.0
Oceania	0	0.0	1	0.0	–	–	0	0.0	–	–	–	–	–	–	0	0.0
Australia	–	–	–	–	–	–	–	–	–	–	–	–	–	–	–	–
New Zealand	0	0.0	0	0.0	–	–	–	–	–	–	–	–	–	–	0	0.0
Other Oceania	–	–	1	0.0	–	–	0	0.0	–	–	–	–	–	–	–	–
North America	184	3.0	293	3.9	329	3.4	74	0.8	–	–	–	–	–	–	–	–
Canada	88	1.4	110	1.5	100	1.0	17	0.2	–	–	–	–	–	–	–	–
United States	96	1.6	183	2.4	224	2.3	57	0.6	–	–	–	–	–	–	–	–
N. America unalloc.	–	–	–	–	4	0.0	–	–	–	–	–	–	–	–	–	–
Other America	205	3.3	109	1.4	118	1.2	145	1.5	173	2.3	232	3.4	400	5.4	430	5.7
Argentina	79	1.3	8	0.1	0	0.0	1	0.0	–	–	–	–	–	–	–	–
Brazil	65	1.1	52	0.7	–	–	31	0.3	–	–	–	–	–	–	–	–
Mexico	–	–	0	0.0	15	0.2	–	–	8	0.1	–	–	–	–	–	–
Venezuela	2	0.0	–	–	0	0.0	–	–	–	–	–	–	–	–	–	–
4 of O.America	146	2.4	60	0.8	16	0.2	31	0.3	8	0.1	–	–	–	–	–	–
Oth.Oth. America	59	1.0	49	0.6	102	1.1	114	1.2	165	2.2	232	3.4	400	5.4	430	5.7
Western Europe	2951	48.1	2920	38.7	4901	50.4	3956	40.3	540	7.3	441	6.4	443	6.0	478	6.3
B.L.E.U.	45	0.7	75	1.0	295	3.0	143	1.5	5	0.1	–	–	5	0.1	7	0.1
Denmark	104	1.7	56	0.7	58	0.6	85	0.9	–	–	4	0.1	8	0.1	1	0.0
France	268	4.4	177	2.3	272	2.8	309	3.1	26	0.3	–	–	0	0.0	19	0.2
Germany,Fed.Rep.	609	9.9	620	8.2	1123	11.5	1017	10.4	–	–	–	–	84	1.1	56	0.7
Ireland	–	–	1	0.0	2	0.0	3	0.0	–	–	–	–	1	0.0	0	0.0
Italy	378	6.2	362	4.8	438	4.5	469	4.8	58	0.8	8	0.1	28	0.4	52	0.7
Netherlands	45	0.7	44	0.6	51	0.5	57	0.6	–	–	–	–	–	–	3	0.0
United Kingdom	123	2.0	80	1.1	118	1.2	138	1.4	43	0.6	–	–	17	0.2	6	0.1
EEC(9)	1571	25.6	1415	18.7	2356	24.2	2220	22.6	132	1.8	12	0.2	143	1.9	145	1.9
Finland	89	1.4	57	0.8	78	0.8	140	1.4	93	1.3	47	0.7	9	0.1	31	0.4
Iceland	6	0.1	2	0.0	2	0.0	3	0.0	1	0.0	4	0.1	1	0.0	–	–
Norway	59	1.0	88	1.2	109	1.1	70	0.7	–	–	–	–	0	0.0	0	0.0
Sweden	94	1.5	113	1.5	99	1.0	140	1.4	2	0.0	1	0.0	2	0.0	0	0.0
Northern Europe	248	4.0	260	3.4	288	3.0	352	3.6	96	1.3	53	0.8	12	0.2	32	0.4
Greece	102	1.7	42	0.6	148	1.5	104	1.1	–	–	–	–	0	0.0	5	0.1
Portugal	5	0.1	1	0.0	12	0.1	33	0.3	–	–	–	–	–	–	0	0.0
Spain	78	1.3	69	0.9	28	0.3	57	0.6	20	0.3	–	–	–	–	0	0.0
Turkey	28	0.5	48	0.6	170	1.7	154	1.6	52	0.7	35	0.5	–	–	–	–
Yugoslavia	614	10.0	704	9.3	1300	13.4	677	6.9	234	3.2	339	4.9	273	3.7	291	3.8
Southern Europe	827	13.5	866	11.5	1657	17.0	1025	10.4	306	4.1	373	5.5	273	3.7	296	3.9
Austria	174	2.8	178	2.4	193	2.0	172	1.8	1	0.0	2	0.0	1	0.0	1	0.0
Switzerland	125	2.0	83	1.1	245	2.5	139	1.4	–	–	–	–	0	0.0	0	0.0
Central Europe	299	4.9	260	3.4	438	4.5	311	3.2	1	0.0	2	0.0	1	0.0	1	0.0
W.Europe unalloc	7	0.1	119	1.6	162	1.7	48	0.5	3	0.0	1	0.0	14	0.2	4	0.1
Eastern Europe	715	11.7	1713	22.7	2029	20.8	2586	26.3	5540	74.8	5200	76.0	6336	86.0	6331	83.6
Albania	.	.	.	.	.	.	.	.								
Bulgaria	.	.	.	.	.	.	.	.	723	9.8	629	9.2	...	...	...	...
Czechoslovakia	.	.	.	.	.	.	.	.	376	5.1	115	1.7	...	...	...	...
German Dem.Rep.	.	.	.	.	.	.	.	.	2599	35.1	2742	40.1	3611	40.9	3352	44.2
Hungary	.	.	.	.	.	.	.	.	547	7.4	543	7.9	...	...	...	...
Poland	.	.	.	.	.	.	.	.	691	9.3	601	8.8	...	...	...	...
Romania	.	.	.	.	.	.	.	.	603	8.1	520	7.6	523	7.1	340	4.5
7 of E.Europe	.	.	.	.	.	.	.	.	5540	74.8	5150	75.3	3534	48.0	3692	48.7
USSR	715	11.7	1713	22.7	2029	20.8	2586	26.3	.	.	.	.				
E.Europe unalloc	.	.	.	.	.	.	.	.	–	–	50	0.7	2802	38.0	2639	34.8
Unallocated	1170	19.1	1640	21.7	1041	10.7	1481	15.1	476	6.4	470	6.9	–	–	–	–
World	6133	100.0	7549	100.0	9733	100.0	9819	100.0	7409	100.0	6839	100.0	7368	100.0	7576	100.0

The east European countries increased their exports to the Far East (from
0.2 million to 0.38 million tonnes) and to the Middle East (from 0.6 million
to 1.13 million tonnes). The share of those exports increased from 3.2 to
3.9 per cent and from 9.6 to 11.5 per cent, respectively.

The share of exports directed to all other regions of the world declined
as follows:

- Africa: from 1.7 to 0.7 per cent;
- Japan: from 0.3 to 0.0 per cent;
- North America: from 3.0 to 0.8 per cent;
- Latin America: from 3.3 to 1.5 per cent;
- EEC: from 25.6 to 22.6 per cent;
- Northern Europe: from 4.0 to 3.6 per cent;
- Southern Europe: from 10.0 to 6.9 per cent; and
- Central Europe: from 4.9 to 3.2 per cent.

The share of east European exports directed to the USSR increased from 11.7 to
26.3 per cent between 1970 and 1982.

Changes in the geographic orientation of USSR steel exports

The steel exports of the USSR remained relatively stable between 1970 and
1982 at some 7.5 million tonnes.

USSR steel exports for the period from 1977 to 1982 have been estimated
according to the methodology described in chapter I. As for the destination
of those exports, the data have been calculated by using the import figures of
some importing countries which report to COMTRADE. As can be seen from
table 2.10, the major part of USSR exports went to the centrally-
planned-economy countries of eastern Europe (approximately 80 per cent). A
small quantity of steel products was exported to the Middle East, the Far East
and Africa.

There was a downward trend in the share of exports to west European
countries; in 1970 it amounted to 7.3 per cent and by 1982 it had dropped to
6.3 per cent.

Changes in the geographic orientation of exports from Oceania

Between 1970 and 1978 there was an upward trend in the exports of steel
from Oceania. They grew from 0.80 million to 2.51 million tonnes. After
that, however, they declined, dropping to 1.26 million tonnes in 1982.

The major part of the Oceanian exports were directed to the Far East.
The share of exports to this region increased from 57.5 to 67.2 per cent
between 1970 and 1982 (table 2.11).

Oceania also increased the share of its exports directed to the
Middle East from 1.4 to 3.7 per cent, to North America from 12.6 to
13.9 per cent and to Latin America from 2.3 to 9.3 per cent.

Oceania virtually stopped all exports of steel products to Japan,
northern Europe and eastern Europe.

The share of exports to EEC and to the south European countries declined
from 12.1 to 2.8 per cent and from 4.7 to 0.8 per cent, respectively.

Table 2.11. Changes in the geographic orientation of exports of steel products

Years / Importers	Oceania 1970 1000t	%	1974 1000t	%	1978 1000t	%	1982 1000t	%	Africa 1970 1000t	%	1974 1000t	%	1978 1000t	%	1982 1000t	%
Africa	11	1.4	11	1.0	13	0.5	2	0.2	.	.	.	.	.	.	.	.
Algeria	-	-	-	-	-	-	-	-	.	.	.	.	.	.	.	.
Liby.Arab Jam.	-	-	1	0.1	-	-	-	-	.	.	.	.	.	.	.	.
Morocco	-	-	-	-	-	-	-	-	.	.	.	.	.	.	.	.
Tunisia	-	-	-	-	-	-	-	-	.	.	.	.	.	.	.	.
Above listed	-	-	1	0.1	-	-	-	-	.	.	.	.	.	.	.	.
South Africa	1	0.1	7	0.7	2	0.1	0	0.0	.	.	.	.	.	.	.	.
Other Africa	10	1.3	3	0.3	12	0.5	2	0.1	.	.	.	.	.	.	.	.
Far East	472	58.9	632	61.2	1517	60.5	848	67.2	5	2.2	9	2.5	286	22.2	88	8.2
China	1	0.1	25	2.4	451	18.0	37	2.9	-	-	-	-	-	-	-	-
India	8	1.0	10	1.0	12	0.5	42	3.4	-	-	-	-	-	-	-	-
Korea,Rep.of	-	-	-	-	154	6.2	121	9.6	-	-	-	-	-	-	-	-
Above listed	9	1.1	35	3.4	618	24.6	200	15.9	-	-	-	-	10	0.8	-	-
Japan	11	1.4	247	24.0	65	2.6	0	0.0	-	-	-	-	10	0.8	-	-
Other Far East	452	56.4	350	33.9	834	33.2	648	51.3	5	2.2	9	2.5	276	21.4	88	8.2
Middle East	11	1.4	20	1.9	189	7.5	47	3.7	26	11.4	132	38.4	36	2.8	0	0.0
Egypt	-	-	-	-	48	1.9	-	-	-	-	-	-	-	-	-	-
Saudi Arabia	-	-	-	-	8	0.3	22	1.8	-	-	-	-	18	1.4	-	-
Above listed	-	-	-	-	56	2.3	22	1.8	-	-	-	-	18	1.4	-	-
Other Middle East	11	1.4	20	1.9	133	5.3	25	1.9	26	11.4	132	38.4	18	1.4	0	0.0
Oceania	.	.	.	.	.	.	.	.	5	2.0	2	0.4	1	0.1	0	0.0
Australia	.	.	.	.	.	.	.	.	0	0.0	-	-	-	-	-	-
New Zealand	.	.	.	.	.	.	.	.	0	0.0	1	0.2	1	0.1	-	-
Other Oceania	.	.	.	.	.	.	.	.	5	2.0	1	0.3	0	0.0	0	0.0
North America	101	12.6	45	4.3	180	7.2	176	13.9	58	26.1	20	5.9	644	49.9	510	47.4
Canada	20	2.5	2	0.2	16	0.6	52	4.1	1	0.3	7	2.0	15	1.2	30	2.8
United States	81	10.1	42	4.1	164	6.5	124	9.8	58	25.7	14	3.9	629	48.7	480	44.7
N. America unalloc.	-	-	0	0.0	0	0.0	-	-								
Other America	18	2.3	53	5.2	261	10.4	117	9.3	22	10.0	33	9.7	99	7.7	64	5.9
Argentina	3	0.4	48	4.7	160	6.4	100	8.0	21	9.4	3	0.8	3	0.3	13	1.2
Brazil	-	-	1	0.1	0	0.0	-	-	-	-	30	8.7	23	1.8	2	0.2
Mexico	-	-	-	-	-	-	0	0.0	-	-	0	0.0	0	0.0	-	-
Venezuela	-	-	0	0.0	55	2.2	-	-	-	-	0	0.0	21	1.6	-	-
Above listed	3	0.4	50	4.6	215	8.6	101	8.0	21	9.4	33	9.6	47	3.6	15	1.4
Oth. Oth. America	15	1.9	4	0.3	46	1.9	16	1.3	1	0.6	0	0.0	52	4.0	49	4.6
Western Europe	160	20.0	251	24.3	346	13.8	45	3.6	108	48.3	148	43.1	225	17.4	414	38.5
B.L.E.U.	7	0.9	80	7.7	78	3.1	34	2.7	0	0.1	1	0.2	9	0.7	34	3.2
Denmark	-	-	-	-	-	-	0	0.0	0	0.0	0	0.1	1	0.1	0	0.0
France	-	-	0	0.0	9	0.3	0	0.0	1	0.2	1	0.2	1	0.1	2	0.1
Germany,Fed.Rep.	11	1.4	170	16.5	63	2.5	0	0.0	10	4.3	62	17.9	44	3.4	139	12.9
Ireland	-	-	0	0.0	0	0.0	-	-	2	0.9	0	0.0	0	0.0	0	0.0
Italy	13	1.6	0	0.0	184	7.3	-	-	34	15.0	24	7.1	27	2.1	19	1.8
Netherlands	-	-	0	0.0	0	0.0	0	0.0	4	2.0	0	0.1	4	0.3	9	0.8
United Kingdom	66	8.2	1	0.1	13	0.5	1	0.0	10	4.7	44	12.9	87	6.7	99	9.2
EEC(9)	97	12.1	251	24.3	346	13.8	35	2.8	61	27.2	132	38.5	173	13.4	303	28.2
Finland	21	2.6	-	-	-	-	-	-	0	0.1	0	0.1	0	0.0	0	0.0
Iceland	-	-	-	-	-	-	-	-								
Norway	-	-	-	-	-	-	-	-	-	-	-	-	-	-	-	-
Sweden	3	0.4	0	0.0	-	-	0	0.0	0	0.0	0	0.1	0	0.0	0	0.0
Northern Europe	24	3.0	0	0.0	0	0.0	0	0.0	0	0.0	0	0.0	0	0.0	0	0.0
Greece	-	-	-	-	-	-	0	0.0	0	0.2	1	0.2	1	0.0	-	-
Portugal	-	-	-	-	-	-	10	0.8	3	1.2	3	0.8	11	0.8	39	3.6
Spain	38	4.7	0	0.0	0	0.0	-	-	0	0.0	10	2.9	10	0.8	16	1.5
Turkey	-	-	-	-	-	-	-	-	43	19.3	2	0.6	16	1.2	39	3.6
Yugoslavia	-	-	-	-	-	-	-	-	-	-	-	-	15	1.1	12	1.1
Southern Europe	38	4.7	0	0.0	0	0.0	10	0.8	46	20.5	15	4.3	51	4.0	105	9.8
Austria	-	-	-	-	-	-	-	-	1	0.4	0	0.0	0	0.0	0	0.0
Switzerland	-	-	0	0.0	0	0.0	-	-	0	0.1	0	0.0	0	0.0	5	0.5
Central Europe	-	-	0	0.0	0	0.0	-	-	1	0.5	0	0.1	0	0.0	6	0.5
W.Europe unalloc	1	0.1	0	0.0	0	0.0										
Eastern Europe	28	3.5	20	1.9	0	0.0	-	-	-	-	-	-	-	-	-	-
Albania	-	-	-	-	-	-										
Bulgaria	-	-	-	-	-	-										
Czechoslovakia	-	-	-	-	-	-										
German Dem.Rep.	-	-	-	-	-	-										
Hungary	-	-	-	-	-	-										
Poland	-	-	20	1.9	-	-										
Romania	28	3.5	-	-	0	0.0										
Above listed	28	3.5	20	1.9	0	0.0										
USSR	-	-	-	-	-	-										
E.Europe unalloc	-	-	-	-	-	-										
Unallocated	-	-	0	0.0	0	0.0	27	2.2	-	-	-	-	-	-	-	-
World	801	100.0	1033	100.0	2507	100.0	1262	100.0	224	100.0	343	100.0	1292	100.0	1075	100.0

Changes in the geographic orientation of steel exports from Africa

The volume of African exports of steel products had shown a continuous increase from 1970 to 1977, when it reached its maximum (1.63 million tonnes) (table 2.11). During this period there were only two significant interruptions in this upward trend - in 1971 (0.1 million tonnes) and in 1975 (0.2 million tonnes). The data presented in table 1 show that, beginning in 1978, the volume of exports started to decline, reaching 0.77 million tonnes in 1982. In 1982 there was a slight increase in exports (1.08 million tonnes).

The major part of African steel exports was directed to North America and to EEC. The growth in the share of those exports between 1970 and 1982 was as follows:

- North America: The share increased from 26.1 to 47.4 per cent;
- EEC: The share increased from 27.2 to 28.2 per cent.

The share of exports to the Far East grew from 2.8 to 8.2 per cent, to southern Europe from 4.3 to 9.8 per cent and to central Europe from 0.1 to 0.5 per cent.

The countries of Africa decreased the share of their exports to the following regions:

- Middle East: from 11.4 per cent to practically zero;
- Oceania: from 2.0 per cent to zero;
- Latin America: from 10.0 to 5.9 per cent;
- Northern Europe: from 0.2 per cent to zero.

Africa exported no steel products to eastern Europe or the USSR.

2.2 Evaluation of the relative importance of each exporter and/or main exporting regions in the various importing countries and/or regions

Africa

The changes occurring between 1970 and 1982 in the shares of steel exports of various exporting regions in the total imports of Africa are presented in table 2.12.

It will be seen that EEC was the main foreign steel supplier of African countries. The share of their exports in total imports of this region fluctuated during the period. In 1970 it accounted for 57.9 per cent, by 1974 it had grown to 63.5 per cent and by 1982 it had dropped to 46.6 per cent.

The South European countries became another principal supplier of African countries. During the period under review, they increased their share in total African steel imports from 1.4 to 23.8 per cent.

While Japan accounted for a fairly considerable proportion of the total steel imports of Africa, it showed a downward trend between 1970 and 1982, with its share decreasing from 23.8 per cent in 1970 to 18.2 per cent in 1982.

The shares of North America and eastern Europe declined from 6.1 to 2.9 per cent and from 3.2 to 1.3 per cent, respectively.

Other exporting regions made up a small proportion of the total imports of steel products of Africa.

Table 2.12. Relative importance of exporting regions
in steel imports of Africa

Years	Africa							
	1970		1974		1978		1982	
Exporters	1 000 t	%	1 000 t	%	1 000 t	%	1 000 t	%
Africa	.	.	.	.	.	.	.	.
Far East	35	1.0	29	0.5	58	1.1	80	1.6
Japan	792	23.8	1 353	24.8	1 054	20.9	931	18.2
Oceania	11	0.3	11	0.2	13	0.3	2	0.0
North America	204	6.1	293	5.4	101	2.0	148	2.9
Latin America	130	3.9	14	0.3	94	1.9	195	3.8
EEC (9)	1 929	57.9	3 466	63.5	2 989	59.2	2 379	46.6
Northern Europe	15	0.5	48	0.9	78	1.5	51	1.0
Southern Europe	46	1.4	106	1.9	498	9.9	1 214	23.8
Central Europe	9	0.3	15	0.3	39	0.8	33	0.7
Eastern Europe	106	3.2	72	1.3	102	2.0	64	1.3
USSR	55	1.6	54	1.0	26	0.5	6	0.1
World	3 332	100.0	5 461	100.0	5 053	100.0	5 103	100.0

Table 2.13. Relative importance of exporting regions
in steel imports of the Far East

Years	Far East							
	1970		1974		1978		1982	
Exporters	1 000 t	%	1 000 t	%	1 000 t	%	1 000 t	%
Africa	5	0.1	9	0.1	286	1.3	88	0.5
Far East	.	.	.	.	.	.	.	.
Japan	5 252	63.4	10 829	76.1	14 055	63.8	12 000	69.0
Oceania	472	5.7	632	4.4	1 517	6.9	848	4.9
North America	836	10.1	605	4.2	555	2.5	699	4.0
Latin America	0	0.0	0	0.0	204	0.9	627	3.6
EEC (9)	1 215	14.7	1 592	11.2	4 287	19.5	2 174	12.5
Northern Europe	60	0.7	41	0.3	174	0.8	103	0.6
Southern Europe	24	0.3	50	0.4	410	1.9	201	1.1
Central Europe	10	0.1	11	0.1	24	0.1	12	0.1
Eastern Europe	213	2.6	252	1.8	420	1.9	384	2.2
USSR	191	2.3	204	1.4	95	0.4	255	1.5
World	8 278	100.0	14 225	100.0	22 026	100.0	17 391	100.0

Far East

In the Far East (see table 2.13), Japan is firmly established as the principal foreign supplier of steel to the region: 76 per cent of total imports in 1974 were provided by Japanese exporters which have the advantage, amongst others, of a shorter transport distance.

In 1970, EEC supplied 14.7 per cent of the import requirements of the Far East but its share had dropped to 12.5 per cent by 1982. While in 1970 the proportion of steel originating from North America reached 10.1 per cent, this was followed by a downward trend and the share in question fell to 4.0 per cent by 1982. The shares of Oceania and eastern Europe in the total steel imports of the Far East also decreased, declining from 5.7 to 4.9 per cent and from 2.6 to 2.2 per cent, respectively, between 1970 and 1982. The share of USSR steel exports dropped from 2.3 to 1.5 per cent between 1970 and 1982.

The following regions increased their shares in total steel imports of the Far East during the period:

- Africa: from 0.1 to 0.5 per cent;
- Latin America: from 0 to 3.6 per cent; and
- Southern Europe: from 0.3 to 1.1 per cent.

The shares of other exporting regions were rather small and underwent no significant change.

Japan

Between 1970 and 1982, the origin of steel exports to Japan showed some modification (see table 2.14). The main foreign steel supplier of Japan was the Far East, with its share of Japanese imports, however, showing a certain fluctuation. In 1970 the share accounted for 54.3 per cent, by 1974 this proportion had dropped to 8.1 per cent and by 1982 it had again grown, attaining 87.3 per cent.

The other main feature of changes which occurred in the origin of steel exports to Japan was the large increase of the Latin American share, which grew from practically zero in 1970 to 11.4 per cent in 1982. Attention must however be drawn to the fact that Japanese imports in volume terms, although growing fast, remain relatively modest when compared with those of other regions.

There was a downward trend in the share of Oceania in the total steel imports of Japan, which, between 1970 and 1982, dropped from 9.4 per cent to zero. The fact that, in 1974, the share reached 82.1 per cent, must be considered an anomaly. The share of steel exports from eastern Europe, which accounted for 13.9 per cent in 1970, increased to 15.1 per cent in 1978 and decreased to zero by 1982.

Other exporting regions, with previously small shares in Japanese imports, saw their shares drop even lower by 1982.

Table 2.14. Relative importance of exporting regions
in steel imports of Japan

Years	Japan							
	1970		1974		1978		1982	
Exporters	1 000 t	%	1 000 t	%	1 000 t	%	1 000 t	%
Africa	0	0.0	0	0.0	0	0.0	0	0.0
Far East	64	54.3	24	8.1	99	26.8	1 694	87.3
Japan	.	.	.	.	.	.	.	.
Oceania	11	9.4	247	82.1	65	17.6	0	0.0
North America	16	13.7	12	4.0	54	14.4	14	0.7
Latin America	0	0.0	0	0.0	26	7.0	221	11.4
EEC (9)	5	4.7	11	3.7	10	2.6	8	0.4
Northern Europe	4	3.7	5	1.5	27	7.2	3	0.2
Southern Europe	–	–	0	0.0	34	9.1	0	0.0
Central Europe	0	0.3	1	0.2	0	0.0	0	0.0
Eastern Europe	16	14.0	1	0.4	56	15.1	–	–
USSR	–	–	–	–	...	...	...	...
World	117	100.0	301	100.0	371	100.0	1 941	100.0

Table 2.15. Relative importance of exporting regions
in steel imports of Latin America

Years	Latin America							
	1970		1974		1978		1982	
Exporters	1 000 t	%	1 000 t	%	1 000 t	%	1 000 t	%
Africa	22	0.5	33	0.3	99	1.4	64	1.3
Far East	2	0.0	118	1.2	7	0.1	102	2.0
Japan	1 566	35.9	4 350	42.6	2 415	34.1	1 478	29.5
Oceania	18	0.4	53	0.5	261	3.7	117	2.3
North America	1 157	26.5	2 172	21.3	956	13.5	757	15.1
Latin America	.	.	.	.	.	.	.	.
EEC (9)	1 152	26.4	2 958	29.0	2 476	35.0	1 622	32.4
Northern Europe	29	0.7	74	0.7	89	1.3	33	0.7
Southern Europe	20	0.5	82	0.8	241	3.4	227	4.5
Central Europe	20	0.5	19	0.2	22	0.3	33	0.7
Eastern Europe	205	4.7	109	1.1	118	1.7	145	2.9
USSR	173	4.0	232	2.3	400	5.6	430	8.6
World	4 365	100.0	10 200	100.0	7 084	100.0	5 008	100.0

Latin America

It may be seen in table 2.15 that EEC was one of the major foreign steel suppliers of Latin American countries. The share of its exports in the total steel imports of that region grew from 26.4 per cent in 1970 to 32.4 per cent in 1982. In 1978, the share was even larger, accounting for 35.0 per cent.

Another main supplier of Latin American countries was Japan, whose share of total steel imports of this region, although fluctuating, accounted for a large proportion between 1970 and 1982. In 1970 it amounted to 35.9 per cent, by 1974 it had grown to 42.6 per cent and by 1982 it had again dropped, to 29.5 per cent.

The share of North America in total steel imports of Latin America showed a downward trend between 1970 and 1982. It decreased from 26.5 to 15.1 per cent.

The shares of other exporting regions recorded the following changes during the period:

- Africa: growth from 0.5 to 1.3 per cent;
- Far East: growth from zero to 2.0 per cent;
- Oceania: growth from 0.4 to 2.3 per cent;
- Northern Europe: stability (0.7 per cent) with some small fluctuations;
- Central Europe: a small increase, from 0.5 to 0.7 per cent;
- Southern Europe: growth from 0.5 to 4.5 per cent;
- Eastern Europe: decline from 4.7 to 2.9 per cent. In 1974 the share was at its lowest, accounting for only 1.1 per cent;
- USSR: decline between 1970 and 1974 from 4.0 to 2.3 per cent. By 1982 it had again grown, to 8.6 per cent.

North America

There were two main foreign suppliers to steel products to North America between 1970 and 1982: Japan and the EEC (see table 2.16). During the period, however, the shares of these two exporting regions dropped from 49.2 to 32.9 per cent and from 43.4 to 37.8 per cent, respectively.

Practically all other exporting regions, with the exception of eastern Europe and the USSR, increased their shares in the total steel imports of North America. Considerable increases were recorded in the shares of Africa (from 0.5 to 4.0 per cent), the Far East (from 0.7 to 7.3 per cent), Latin America (from 2.3 to 8.0 per cent) and southern Europe (from 0.5 to 4.1 per cent).

Middle East

Between 1970 and 1982, the principal supplier of foreign steel products to the region was Japan (see table 2.17). The share of Japanese steel in total steel imports of the Middle East increased from 19.5 per cent in 1970 to 41.3 per cent in 1982. It was not, however, a linear increase. In 1974 it accounted for 42.2 per cent, by 1978 it had dropped to 35.3 per cent and by 1982 it had again grown, to 41.3 per cent.

Table 2.16.   Relative importance of exporting regions
in steel imports of North America

	North America							
	1970		1974		1978		1982	
Years								
Exporters	1 000 t	%	1 000 t	%	1 000 t	%	1 000 t	%
Africa	58	0.5	20	0.1	644	3.8	510	4.0
Far East	87	0.7	858	5.7	1 058	6.3	924	7.3
Japan	5 907	49.2	6 675	44.1	5 996	35.6	4 170	32.9
Oceania	101	0.8	45	0.3	180	1.1	176	1.4
North America	.	.	.	.	.	.	.	.
Latin America	277	2.3	257	1.7	736	4.4	1 021	8.0
EEC (9)	5 214	43.4	6 649	43.9	6 786	40.3	4 798	37.8
Northern Europe	87	0.7	158	1.0	433	2.6	418	3.3
Southern Europe	63	0.5	145	1.0	656	3.9	524	4.1
Central Europe	31	0.3	35	0.2	24	0.1	74	0.6
Eastern Europe	184	1.5	293	1.9	329	2.0	74	0.6
USSR	–	–	–	–	...	...	...	...
World	12 009	100.0	15 134	100.0	16 841	100.0	12 688	100.0

Table 2.17.   Relative importance of exporting regions
in steel imports of the Middle East

	Middle East							
	1970		1974		1978		1982	
Years								
Exporters	1 000 t	%	1 000 t	%	1 000 t	%	1 000 t	%
Africa	26	0.8	132	1.7	36	0.3	0	0.0
Far East	277	8.9	149	1.9	501	4.7	1 036	7.4
Japan	608	19.5	3 223	42.2	3 733	35.3	5 810	41.3
Oceania	11	0.4	20	0.3	189	1.8	47	0.3
North America	72	2.3	280	3.7	199	1.9	629	4.5
Latin America	8	0.3	5	0.1	50	0.5	529	3.8
EEC (9)	1 010	32.4	2 768	36.2	4 175	39.5	2 609	18.6
Northern Europe	12	0.4	36	0.5	119	1.1	130	0.9
Southern Europe	36	1.1	175	2.3	673	6.4	1 947	13.8
Central Europe	31	1.0	72	0.9	31	0.3	120	0.9
Eastern Europe	589	18.9	549	7.2	793	7.5	1 128	8.0
USSR	435	14.0	238	3.1	68	0.6	75	0.5
World	3 116	100.0	7 646	100.0	10 568	100.0	14 061	100.0

In 1970, EEC had the biggest share in the region's imports amounting to 32.4 per cent. After that, the situation changed, with the share growing to 1978 to attain 39.5 per cent and, by 1982, dropping to 18.6 per cent.

Southern Europe considerably increased its share in the total steel imports of the region, from 1.1 to 13.8 per cent betwen 1970 and 1982. Latin America and North America also increased their shares, from 0.3 to 3.8 per cent and from 2.3 to 4.5 per cent, respectively. The share of northern Europe increased from 0.4 to 0.9 per cent.

The share of imports from Oceania remained more or less stable, accounting for 0.3 per cent in 1982. The share of central Europe, at around 0.9 per cent, showed no significant change during the period.

The shares of other exporting regions declined between 1970 and 1982:

- Africa: from 0.8 per cent to 0;
- Far East: from 8.9 to 7.4 per cent. In 1974 it accounted for only 1.9 per cent;
- Eastern Europe: from 18.9 to 8.0 per cent; and
- USSR: from 14.0 to 0.5 per cent.

Oceania

During the period under review, as shown in table 2.18, the major supplier of steel products to Oceania was Japan. The share of Japanese steel in the total imports of Oceania grew from 67.2 per cent in 1970 to 78.7 per cent in 1974, declining to 75.1 per cent in 1978 and, by 1982, increasing again, to 77.0 per cent.

In 1970, about 24 per cent of the total steel imports of Oceania came from EEC member countries. Between 1970 and 1982, this share dropped to 10.6 per cent.

The Far East recorded a significant increase in its share in the region's steel imports. It grew from 0.8 to 9.3 per cent between 1970 and 1978 and went on to decrease slightly by 1982, dropping to 8.5 per cent.

Latin American countries started exporting steel products to Oceania only in the second half of the 1970s. By 1982 their share of these exports had grown to 1.8 per cent.

Eastern Europe and the USSR did not export steel products to Oceania. The share of North America and Africa in the total steel imports of Oceania decreased, between 1970 and 1982, from 6.3 to 1.2 per cent, and from 0.5 per cent to zero, respectively. The shares of all other exporting regions were insignificant and remained virtually unchanged.

European Economic Community

As may be seen in table 2.19, the major supplier of steel products to EEC in 1970 was North America. Its share in the total imports of the region accounted for 34.9 per cent. Between 1970 and 1982 North America lost that position and, by 1982, its share amounted to only 4.8 per cent.

Table 2.18.   Relative importance of exporting regions
in steel imports of Oceania

	Oceania							
Years	1970		1974		1978		1982	
Exporters	1 000 t	%	1 000 t	%	1 000 t	%	1 000 t	%
Africa	5	0.5	2	0.1	1	0.2	0	0.0
Far East	8	0.8	70	4.4	76	9.3	103	8.5
Japan	594	67.2	1 248	78.7	618	75.1	940	77.0
Oceania	.	.	.	.	.	.	.	.
North America	55	6.3	67	4.2	17	2.1	15	1.2
Latin America	0	0.0	0	0.0	1	0.1	22	1.8
EEC (9)	210	23.7	182	11.5	97	11.8	129	10.6
Northern Europe	10	1.1	12	0.8	7	0.8	7	0.6
Southern Europe	1	0.1	0	0.0	3	0.4	1	0.1
Central Europe	3	0.3	4	0.2	2	0.2	3	0.2
Eastern Europe	-	-	1	0.1	-	-	-	-
USSR	-	-	-	-	...	...	...	...
World	885	100.0	1 586	100.0	822	100.0	1 221	100.0

Table 2.19.   Relative importance of exporting regions in steel
imports of the European Economic Community

	European Economic Community							
Years	1970		1974		1978		1982	
Exporters	1 000 t	%	1 000 t	%	1 000 t	%	1 000 t	%
Africa	61	0.7	132	2.1	173	1.9	303	3.3
Far East	23	0.3	38	0.6	86	1.0	145	1.6
Japan	1 038	12.2	1 054	16.4	623	6.9	260	2.9
Oceania	97	1.1	251	3.9	346	3.8	35	0.4
North America	2 958	34.9	497	7.7	264	2.9	430	4.8
Latin America	80	0.9	16	0.2	152	1.7	589	6.5
EEC (9)	.	.	.	.	.	.	.	.
Northern Europe	1 421	16.8	1 567	24.4	2 085	23.1	1 994	22.1
Southern Europe	255	3.0	434	6.8	1 177	13.0	1 156	12.8
Central Europe	844	10.0	1 005	15.7	1 626	18.0	1 764	19.5
Eastern Europe	1 571	18.5	1 415	22.0	2 356	26.1	2 220	24.6
USSR	132	1.6	12	0.2	143	1.6	145	1.6
World	8 481	100.0	6 421	100.0	9 031	100.0	9 040	100.0

The share of Japanese steel exports grew from 12.2 per cent in 1970 to 16.4 per cent in 1974. By 1982 it had dropped to 2.9 per cent.

By 1982 the other west European and the east European countries had become the principal suppliers of steel products to EEC. North European countries increased their share in total EEC steel imports from 16.8 per cent in 1970 to 22.1 per cent in 1982, South European countries from 3.0 to 12.8 per cent, Central European countries from 10.0 to 19.5 per cent and east European countries from 18.5 to 24.6 per cent.

The share of Latin American countries in the total EEC steel imports grew from 0.9 to 6.5 per cent between 1970 and 1982. The shares of steel exports from Africa and the Far East also recorded an increase, from 0.7 to 3.3 per cent and from 0.3 to 1.6 per cent, respectively.

The USSR share in EEC steel imports was fairly stable between 1970 and 1982, ranging around 1.6 per cent.

The share of Oceanian steel exports in total EEC imports was insignificant and actually declined between 1970 and 1982.

Northern countries of western Europe

Between 1970 and 1982, as shown in table 2.20 EEC was the principal steel supplier of the region. The share of its steel exports increased from 77.2 to 78.7 per cent between 1970 and 1974 and, between then and 1982, dropped to 58.3 per cent.

The Far East rapidly improved its position as a steel exporter to the region. Its exports to northern Europe, which commenced only at the end of the 1970s, saw their share grow from 0.2 per cent in 1978 to 15.6 per cent in 1982.

The share of Japanese steel exports fluctuated between 1970 and 1982. It accounted for 6.2 per cent in 1970, 9.6 per cent in 1974, 4.4 per cent in 1978 and 6.9 per cent in 1982.

The share of the east European countries' steel exports also fluctuated. It amounted to 7.6 per cent in 1970, 6.3 per cent in 1974, 10.7 per cent in 1978 and 9.3 per cent in 1982.

Between 1970 and 1978 the shares of southern and central countries of western Europe grew from 0.3 to 6.7 per cent and from 2.0 to 3.9 per cent, respectively. By 1982 they had, however, again dropped below the 1978 level.

The shares of steel exports from North America and the USSR in total imports of the region declined during the period, that of North America from 2.8 to 0.4 per cent and that of the USSR from 3.0 to 0.8 per cent.

Latin American countries exported only small quantities of steel products to the region. In 1970 their share accounted for 0.3 per cent and by 1982 it had grown to 0.6 per cent. Africa and Oceania recorded practically no steel exports to the region.

Table 2.20, Relative importance of exporting regions in steel imports of northern countries of western Europe

	Northern countries of western Europe							
Years	1970		1974		1978		1982	
Exporters	1 000 t	%	1 000 t	%	1 000 t	%	1 000 t	%
Africa	0	0.0	1	0.0	1	0.0	0	0.0
Far East	0	0.0	0	0.0	7	0.2	589	15.6
Japan	201	6.2	394	9.6	118	4.4	262	6.9
Oceania	24	0.7	0	0.0	0	0.0	0	0.0
North America	90	2.8	39	0.9	7	0.3	16	0.4
Latin America	10	0.3	1	0.0	0	0.0	22	0.6
EEC (9)	2 521	77.2	3 230	78.7	1 980	73.3	2 205	58.3
Northern Europe	.	.	.	.	.	.	.	.
Southern Europe	9	0.3	13	0.3	181	6.7	195	5.2
Central Europe	65	2.0	112	2.7	105	3.9	108	2.9
Eastern Europe	248	7.6	260	6.3	288	10.7	352	9.3
USSR	96	3.0	53	1.3	12	0.5	32	0.8
World	3 264	100.0	4 102	100.0	2 700	100.0	3 781	100.0

Table 2.21, Relative importance of exporting regions in steel imports of southern countries of western Europe

	Southern countries of western Europe							
Years	1970		1974		1978		1982	
Exporters	1 000 t	%	1 000 t	%	1 000 t	%	1 000 t	%
Africa	46	0.9	15	0.2	51	0.9	105	1.8
Far East	11	0.2	20	0.3	1	0.0	24	0.4
Japan	1 027	20.5	1 333	20.6	516	9.6	312	5.4
Oceania	38	0.8	0	0.0	0	0.0	10	0.2
North America	583	11.6	228	3.5	39	0.7	226	3.9
Latin America	39	0.8	4	0.1	171	3.2	78	1.4
EEC (9)	2 006	40.0	3 380	52.3	2 447	45.5	3 352	58.5
Northern Europe	75	1.5	111	1.7	42	0.8	167	2.9
Southern Europe	.	.	.	.	.	.	.	.
Central Europe	55	1.1	127	2.0	185	3.4	138	2.4
Eastern Europe	827	16.5	866	13.4	1 657	30.8	1 025	17.9
USSR	306	6.1	373	5.8	273	5.1	296	5.2
World	5 013	100.0	6 457	100.0	5 382	100.0	5 734	100.0

Southern countries of western Europe

As indicated in table 2.21, EEC occupied the first position amongst the steel exporters to the region. Its share in total imports increased from 40.0 per cent in 1970 to 58.3 per cent in 1982.

In 1970 Japan was the second exporter of steel products to the region with its share in total imports accounting for 20.5 per cent. This was followed by a downward trend and by 1982 its share had dropped to 5.4 per cent.

The share of the east European countries' exports of steel products grew from 16.5 per cent in 1970 to 30.8 per cent in 1978 and by 1982 had dropped to 17.9 per cent.

The shares of other steel exporting regions in the total steel imports of the southern countries of western Europe recorded the following movements:

- Africa: growth from 0.9 to 1.8 per cent;
- Far East: growth from 0.2 to 0.4 per cent;
- Oceania: decline from 0.8 to 0.2 per cent;
- North America: decline from 11.6 to 0.7 per cent in 1978; by 1982 it had grown to 3.9 per cent;
- Latin America: growth from 0.8 to 3.2 per cent in 1978; by 1982 it had dropped to 1.4 per cent;
- USSR: decline from 6.1 to 5.2 per cent.

Central countries of western Europe

Between 1970 and 1982, EEC was the major foreign supplier of steel products to the region (see table 2.22). Its share in total steel imports grew from 81.3 per cent in 1970 to 83.8 per cent in 1974; by 1982 it had dropped to 76.4 per cent.

The remaining part of the region's steel imports were provided mainly by east European countries and the northern and southern countries of western Europe. The shares of steel exports from these regions in the total steel imports of the central countries of western Europe recorded the following changes:

- Eastern Europe: the share fluctuated at around 13 per cent;
- Northern Europe: growth from 1.8 to 2.2 per cent;
- Southern Europe : growth from 0.5 to 6.2 per cent.

The share of North American countries dropped significantly, from 3.2 to 0.9 per cent. As seen in table 2.22, other steel-exporting regions accounted for only a small proportion in total steel imports of the region with their shares remaining virtually stable.

East European countries

The changes in the shares of the main steel-exporting regions in the total steel imports of the east European countries are shown in table 2.23. The data concerning USSR steel exports were estimated using the methodology described in Chapter I.

Table 2.22. Relative importance of exporting regions in steel
imports of central countries of western Europe

	Central countries of western Europe							
Years	1970		1974		1978		1982	
Exporters	1 000 t	%	1 000 t	%	1 000 t	%	1 000 t	%
Africa	1	0.0	0	0.0	0	0.0	6	0.2
Far East	-	-	0	0.0	1	0.0	2	0.1
Japan	11	0.5	63	2.6	85	3.1	19	0.8
Oceania	-	-	0	0.0	-	-	-	-
North America	75	3.2	6	0.2	8	0.3	21	0.9
Latin America	0	0.0	-	-	-	-	13	0.5
EEC (9)	1 907	81.3	2 019	83.8	1 966	72.6	1 864	76.4
Northern Europe	42	1.8	39	1.6	54	2.0	53	2.2
Southern Europe	11	0.5	21	0.9	153	5.7	151	6.2
Central Europe	.	.	.	.	.	.	.	.
Eastern Europe	299	12.7	260	10.8	438	16.2	311	12.7
USSR	1	0.0	2	0.1	1	0.0	1	0.0
World	2 346	100.0	2 410	100.0	2 706	100.0	2 440	100.0

Table 2.23. Relative importance of exporting regions in steel
imports of east European countries

	East European countries							
Years	1970		1974		1978		1982	
Exporters	1 000 t	%	1 000 t	%	1 000 t	%	1 000 t	%
Africa	-	-	-	-	-	-	-	-
Far East	2	0.0	2	0.0	0	0.0	-	-
Japan	248	3.6	259	3.0	148	1.7	41	0.6
Oceania	28	0.4	20	0.2	0	0.0	-	-
North America	40	0.6	41	0.5	32	0.4	1	0.0
Latin America	0	0.0	0	0.0	2	0.0	0	0.0
EEC (9)	782	11.3	2 421	28.2	1 574	18.4	438	6.1
Northern Europe	44	0.6	158	1.9	87	1.0	41	0.6
Southern Europe	120	1.7	281	3.3	235	2.8	228	3.2
Central Europe	138	2.0	197	2.3	143	1.7	74	1.0
Eastern Europe	.	.	.	.	.	.	.	.
USSR	5 540	79.8	5 200	60.6	6 336	74.0	6 331	88.5
World	6 941	100.0	8 579	100.0	8 557	100.0	7 154	100.0

As can be seen from the table, the USSR was the principal supplier of foreign produced steel products to the region: in 1970, close to 80 per cent of the total imports were provided by the USSR. Although by 1974 its share had dropped to 60.6 per cent, there was subsequently an upward trend when the share increased to 88.5 per cent by 1982.

Between 1970 and 1974, the share of steel exports from EEC grew from 11.3 to 28.2 per cent; by 1982 it had, however, dropped to 6.1 per cent.

During the period under review, only the southern countries of western Europe increased their share in the total imports of the region, growing from 1.7 to 3.2 per cent.

North America and Latin America recorded practically no exports of steel products to the region. Africa did not export steel to the region.

The small share of the other exporting regions in the total imports of the east European countries declined even further betwen 1970 and 1982.

USSR

Changes occurring between 1970 and 1982 in the shares of the main exporting regions in total steel imports of the USSR are shown in table 2.24. The data on USSR steel imports from the east European countries for 1978 and 1982 were estimated using the methodology described in Chapter I.

EEC was the major supplier of foreign steel products to the USSR. In 1970 its share accounted for 42.6 per cent and by 1974 it had attained 54.5 per cent. After 1974 there was a downward trend, with the share dropping to 41.7 per cent by 1982.

The east European countries occupied the second position amongst steel exporters to the USSR, in spite of the fact that their share dropped from 31.8 to 25.7 per cent between 1970 and 1982.

Japan considerably increased its share in USSR steel imports. Between 1970 and 1982 it grew from 9.7 to 23.6 per cent.

The share of the southern countries of western Europe fluctuated during the period. In 1970 it accounted for 1.4 per cent and in 1978 for 4.9 per cent. By 1982 it had dropped to 2.7 per cent. The share of the central countries of western Europe accounted for 4.4 per cent in 1970, dropping to 2.0 per cent by 1974 and, by 1982, again growing, to 5.3 per cent. The share of the northern countries of western Europe in total imports of the USSR decreased, with some fluctuations, from 1.9 to 0.7 per cent between 1970 and 1982.

After recording a share of 2.1 per cent in 1970, the North American countries practically stopped exporting steel products to the USSR. The share of the Far East dropped from 6.1 per cent in 1970 to 0.2 per cent in 1982. The share of Latin American countries was insignificant, accounting for only 0.2 per cent in 1982.

Africa and Oceania did not export steel products to the USSR.

Table 2.24.   Relative importance of exporting regions
in steel imports of the USSR

Years	USSR							
	1970		1974		1978		1982	
Exporters	1 000 t	%	1 000 t	%	1 000 t	%	1 000 t	%
Africa	-	-	-	-	-	-	-	-
Far East	137	6.1	0	0.0	86	1.0	15	0.2
Japan	218	9.7	1 317	17.3	1 516	17.0	2 380	23.6
Oceania	-	-	-	-	-	-	-	-
North America	48	2.1	24	0.3	8	0.1	2	0.0
Latin America	-	-	1	0.0	13	0.1	16	0.2
EEC (9)	958	42.6	4 149	54.5	4 466	50.0	4 196	41.7
Northern Europe	43	1.9	34	0.5	106	1.2	75	0.1
Southern Europe	32	1.4	228	3.0	437	4.9	272	2.7
Central Europe	98	4.4	153	2.0	262	2.9	529	5.3
Eastern Europe	715	31.8	1 713	22.5	2 029	22.7	2 586	25.7
USSR	.	.	.	.	.	.	.	.
World	2 250	100.0	7 619	100.0	8 924	100.0	10 072	100.0

2.3   Examination of structural changes in total deliveries of steel by type of steel product being exported by various countries and/or regions (without analysis of their destination)

The product composition of world trade in steel has for some time been moving towards more highly finished steel, particularly flat-rolled products, as well as high-quality steels. When analysed at world level, however, this trend may be somewhat attenuated by the broad level of aggregation used in the statistics.

Part A of table 2.25 presents the volume and the share of particular product groups in the world trade in steel of the various regions. It should be mentioned that no account has been taken of the exports of those countries which have not provided these data broken down by product.

The group of flat products accounted for the largest proportion of total world steel exports. However, between 1970 and 1982, the share of flat products in world steel trade decreased from 43.6 to 36.8 per cent.

The data presented illustrate the importance of the rather considerable proportion of long products. The share of long products in world trade in steel increased from 25.7 to 28.5 per cent during the period. This may be explained by the fact that this group includes very high quality products which enter international trade for exchange purposes, reflecting the comparative advantages of various steel producing countries.

One of the most striking developments is the growth of the group of tubes and fittings. This is, to some extent, due to the increased imports of this type of steel product by oil-producing countries, eastern Europe and the USSR. The share of these products grew from 12.0 to 19.1 per cent between 1970 and 1982.

Table 2.25 shows the relatively high proportion of the group of ingots and semis in steel exports. It should be borne in mind that this group comprises many products which, although semi-finished, are comparatively specialized. This is due in part to the growing practice of international re-rolling contracts. The share of this group of steel products declined from 16.4 per cent in 1970 to 14.0 per cent in 1982.

A rather low level may be noted in the proportion of wire, which can be produced more easily from imported rods. The share of wire in world trade in steel products declined from 2.3 per cent to 1.6 per cent during the period.

The European Economic Community

The changes which took place in the product pattern of EEC exports are presented in part B of table 2.25. The main feature of the product pattern of EEC steel trade is the decline of the groups of long products and flat products, which in 1970 accounted for, respectively, 32.2 per cent and 41.1 per cent of total EEC trade in steel products. During the period under review, their shares had decreased, respectively, to 28.6 per cent and to 32.8 per cent.

The other trend brought out by the table is the relative growth of the share of the group of ingots and semis, which can be partially explained by the increasing shipments of coils of flat products. The share of this group of products grew from 9.1 per cent in 1970 to 18.1 per cent in 1982.

The share of the group of tubes and fittings increased considerably between 1970 and 1982 (from 14.4 to 23.4 per cent).

The group of wire accounted for only a small proportion of the total EEC exports of steel products, with a share declining slightly from 3.1 to 1.9 per cent during the period.

North European countries

Part C of table 2.25 shows the product composition of exports from the North European countries and the changes in the product pattern which occurred from 1970 to 1982.

The most important development was the advance of the group of flat products. The share of this group of products in total steel exports of Northern Europe increased from 32.8 to 43.0 per cent between 1970 and 1982. This development was due in particular to the specialization of some countries in the production of certain types of products as well as to the destination of their exports.

Another important group of products was long products. The share of these decreased slightly, from 28.6 to 27.0 per cent.

A considerable part of the North European countries' exports was exported in the form of ingots and semis. Their share accounted for 24.8 per cent in 1970 and then declined gradually, amounting to only 4.5 per cent by 1982.

As in all main regions, the group of wire constituted the smallest category of exported products, its share having dropped from 3.6 to 1.9 per cent between 1970 and 1982.

The share of tubes and fittings decreased from 9.8 to 7.8 per cent.

South European countries

The changes occurring in the structure of steel product exports of the South European countries are shown in part D of table 2.25.

The major part of the exports of this region was made up of the group of long products. The share of long products in the total steel exports of the South European countries increased considerably - from 41.5 to 60.1 per cent - between 1970 and 1982.

The share of each of the other four groups of products decreased in the following way during the same period:

- Ingots and semis: from 22.8 to 11.5 per cent;
- Flat products: from 18.2 to 15.2 per cent;
- Tubes and fittings: from 16.3 to 12.1 per cent;
- Wire: from 1.2 to 1.1 per cent.

Central European countries

Part E of table 2.25 brings out the point that, in 1982, the group of flat products represented 49.3 per cent or almost half of total steel exports of those countries compared with 41.2 per cent in 1970.

A considerable falling off may be noted in the proportion of ingots and semis, whose share decreased from 28.7 per cent in 1970 to 9.5 per cent in 1982.

The other major trend brought out by the table is the relative growth over the period under review in the proportion of long products. The share of long products grew from 17.5 to 23.7 per cent.

An increase occurred for the group of tubes and fittings, whose share grew from 9.0 to 15.2 per cent.

The smallest portion was accounted for by wire. As in the other regions, between 1970 and 1982, there was a downward trend in the share of wire in total steel exports of the Central European countries. Its share decreased from 3.6 to 2.3 per cent.

East European countries

During the period under review, the following changes occurred in the structure of the steel exports of the east European countries (part F of table 2.25):

- In 1982 the share of ingots and semis decreased by 1.5 per cent compared with 1970 (from 10.4 to 8.9 per cent);

- The share of long products in steel exports accounted for 39.8 per cent in 1982, i.e. the main part of their steel exports. There was a slight increase (of 3.0 per cent) compared with 1970;

- The share of flat products has not changed significantly between 1970 and 1982 (from 36.8 to 35.8 per cent);

- The share of wire more than doubled during the same period (from 1.6 to 3.8 per cent);

- While the group of tubes and fittings occupied the third position after the group of long products and flat products, their share declined from 14.5 to 11.8 per cent.

Union of Soviet Socialist Republics

Owing to the lack of information from the Soviet Union concerning its exports by categories of steel products, it was impossible to analyse the structural changes in product composition of steel exports for the whole period, the required statistical data being available for the period from 1970 to 1975 only. However, by considering the pattern of the USSR steel exports in value terms, it was possible to assume that, globally, during the period from 1976 to 1982, there were no significant changes in the product pattern of its exports.

Between 1970 and 1974 the structure of the USSR exports of steel products changed in the following way (part G of table 2.25):

- The share of ingots and semis decreased from 16.4 to 15.8 per cent;

- The share of long products grew from 35.6 to 37.7 per cent;

- The share of flat products decreased from 42.7 to 40.9 per cent;

- The share of wire and tubes and fittings increased slightly, from 1.0 to 1.1 per cent and from 4.3 to 4.4 per cent, respectively.

Japan

The changes which took place in the product pattern of Japanese exports of steel products are presented in part H of table 2.25.

The major part of Japanese steel exports was made up of the group of flat products. But during the period under review, the share of direct exports of this group of products in total Japanese steel exports declined from 57 per cent in 1970 to 37.9 per cent in 1982.

Of Japanese exports, 28.2 per cent were accounted for by the group of long products, which totalled 13.7 per cent in 1970. Such a high proportion of long products could be partly explained by the large share of the country's exports directed to developing countries.

Another important group of products in Japanese exports was tubes and fittings. The share of these products increased from 15.2 to 24.1 per cent between 1970 and 1982. These figures bear out the emergence of Japan as the main international steel-pipe supplier during the period 1970-1982.

The shares of ingots and semis and wire varied relatively little over the period, decreasing from 11.4 to 8.9 per cent and from 2.7 to 0.9 per cent, respectively.

Far Eastern countries

There were significant structural changes in the product pattern of the total exports of this region between 1970 and 1982 (part I of table 2.25).

The main feature of the changes was the decline in the share of long products and the increase in that of flat products. The share of long products, which accounted for 73.9 per cent in 1970, decreased to 17.7 per cent in 1982. The share of flat products in total exports grew from 6.5 per cent to 38.8 per cent.

At the same time the shares of ingots and semis and of tubes and fittings increased from 7.6 to 26.9 per cent and from 11.7 to 16.4 per cent, respectively. The share of wire in total exports of this region was rather small fluctuating between 0.2 and 1.3 per cent.

North American countries

Contrary to all other regions, the volume of steel exports of North America decreased strongly between 1970 and 1982 (6.1 Mt to 2.9 Mt). All product categories underwent a strong reduction, with ingots and semis showing the strongest decline (3.0 Mt in 1970, 0.4 Mt in 1982). This evolution deeply modified the structure of the region's steel exports. The reduction of the relative share of ingots and semis from 49.1 per cent in 1970 to 14.2 per cent in 1982 brought about an increase of the relative share of all the other product categories (part J of table 2.25).

Steel product exports from the North American countries showed the following further changes during the period:

- The share of long products grew from 11.0 to 15.9 per cent;

- A major increase was achieved in the proportion of flat products, whose share increased from 34.4 to 56.5 per cent;

- There was a slight increase in the share of wire (from 0.4 to 0.6 per cent);

- The share of tubes and fittings more than doubled (from 5.1 to 12.8 per cent).

Latin American countries

The significance of the group of long products in total exports of the Latin American countries decreased, their share changing from 45.9 per cent to 28.8 per cent between 1970 and 1982 (part K of table 2.25).

The progression of flat products was particularly notable, as their proportion grew from 0.2 Mt in 1970 (41.4 per cent of steel exports) to 1.6 Mt (51.4 per cent) in 1982. During the same period the shares of tubes and fittings and of ingots and semis increased from 8.5 to 11.5 per cent and from 3.5 to 7.9 per cent, respectively. The share of wire declined from 0.7 to 0.3 per cent.

Oceania

The changes observed in the product pattern of the exports from the countries of Oceania are the rising share of long products (from 4.5 to 11.7 per cent) and of flat products (from 32.4 to 34.3 per cent) (part L of table 2.25).

The major share of the total steel exports of this region, accounting for 77.1 per cent in 1974, was made up of the group of ingots and semis. But after 1974 there was a downward trend and by 1982 the share of these products decreased to 51.0 per cent.

The shares of wire and tubes and fittings decreased over the period from 3.0 to 0.7 per cent and from 4.6 to 2.3 per cent, respectively.

Africa

There were considerable structural changes in African steel exports during the period. The main development was the decrease in the share of ingots and semis from 54.6 per cent in 1970 to 20.0 per cent in 1982, and the growth of flat products from 9.9 to 35.1 per cent (part M of table 2.25).

The increase in the share of long products was not significant. Between 1970 and 1982 it grew from 35.2 to 37.5 per cent (it reached 44.8 per cent in 1978).

The share of wire in total steel exports of this region showed a slight increase, from 0.3 to 0.8 per cent.

The share of tubes and fittings increased from 0.1 to 6.8 per cent between 1970 and 1982.

Table 2.25. Structural changes in the exports of steel products

(In calculating the shares of particular product groups in steel exports for tables A and I, no account was taken of the exports of those countries which did not provide these data broken down by product)

A. World

Products	1970 1000t	%	1974 1000t	%	1978 1000t	%	1982 1000t	%
Ingots and semis	10 193	16.4	14 919	16.2	16 338	17.4	12 548	14.0
Long products	15 943	25.7	27 447	29.8	24 771	26.4	25 552	28.5
Flat products	27 083	43.6	36 763	39.8	35 397	37.8	33 042	36.8
Wire	1 410	2.3	1 822	2.0	1 705	1.8	1 378	1.6
Tubes and fittings	7 476	12.0	11 287	12.2	15 596	16.6	17 165	19.1
Total	62 450	100.0	92 436	100.0	102 069		99 110	

B. European Economic Community

Products	1970 1000t	%	1974 1000t	%	1978 1000t	%	1982 1000t	%
Ingots and semis	1 737	9.1	2 421	7.4	5 182	15.5	4 683	18.1
Long products	6 121	32.2	12 375	37.6	9 401	28.2	5 600	21.7
Flat products	7 798	41.1	12 996	39.5	12 351	37.1	9 009	34.9
Wire	594	3.1	853	2.6	602	1.8	499	1.9
Tubes and fittings	2 742	14.4	4 291	13.0	5 799	17.4	6 039	23.4
Total	18 992	100.0	32 936	100.0	33 335	100.0	25 831	100.0

C. North European Countries

Products	1970 1000t	%	1974 1000t	%	1978 1000t	%	1982 1000t	%
Ingots and semis	456	24.8	508	22.3	866	26.4	623	20.3
Long products	526	28.6	676	29.6	858	26.2	829	27.0
Flat products	602	32.8	791	34.7	1 218	37.2	1 322	43.0
Wire	63	3.4	83	3.6	73	2.2	59	1.9
Tubes and fittings	191	10.4	223	9.8	262	8.0	240	7.8
Total	1 838	100.0	2 281	100.0	3 277	100.0	3 073	100.0

Table 2.25. Structural changes in the exports of steel products (continued)

Years / Products	D. South European Countries							
	1970		1974		1978		1982	
	1000t	%	1000t	%	1000t	%	1000t	%
Ingots and semis	141	22.8	75	4.8	505	10.7	702	11.5
Long products	258	41.5	791	51.1	2 458	51.9	3 683	60.1
Flat products	113	18.2	383	24.8	1 114	23.5	930	15.2
Wire	7	1.2	13	0.8	33	0.7	70	1.1
Tubes and fittings	101	16.3	285	18.5	622	13.1	742	12.1
Total	621	100.0	1 547	100.0	4 732	100.0	6 127	100.0

Years / Products	E. Central European Countries							
	1970		1974		1978		1982	
	1000t	%	1000t	%	1000t	%	1000t	%
Ingots and semis	375	28.7	387	22.1	454	18.4	275	9.5
Long products	228	17.5	317	18.1	546	22.2	684	23.7
Flat products	537	41.2	752	43.0	1 135	46.1	1 423	49.3
Wire	47	3.6	59	3.4	60	2.4	66	2.3
Tubes and fittings	117	9.0	234	13.4	268	10.0	440	15.2
Total	1 304	100.0	1 749	100.0	2 463	100.0	2 889	100.0

Years / Products	F. East European Countries							
	1970		1974		1978		1982	
	1000t	%	1000t	%	1000t	%	1000t	%
Ingots and semis	639	10.4	915	12.1	2 298	23.6	872	8.9
Long products	2 256	36.8	2 270	30.1	2 697	27.7	3 906	39.8
Flat products	2 254	36.8	3 295	43.6	3 158	32.5	3 510	35.8
Wire	96	1.6	176	2.3	482	4.9	372	3.8
Tubes and fittings	888	14.5	893	11.8	1 098	11.3	1 158	11.8
Total	6 133	100.0	7 549	100.0	9 733	100.0	9 819	100.0

Table 2.25. Structural changes in the exports of steel products (continued)

G. USSR

Years	1970		1974		1978		1982	
Products	1000t	%	1000t	%	1000t	%	1000t	%
Ingots and semis	1 214	16.4	1 080	15.8	-	-	-	-
Long products	2 636	35.6	2 581	37.7	-	-	-	-
Flat products	3 162	42.7	2 799	40.9	-	-	-	-
Wire	77	1.0	77	1.1	-	-	-	-
Tubes and fittings	321	4.3	302	4.4	-	-	-	-
Total	7 409	100.0	6 839	100.0	7 368	100.0	7 575	100.0

H. Japan

Years	1970		1974		1978		1982	
Products	1000t	%	1000t	%	1000t	%	1000t	%
Ingots and semis	1 990	11.4	7 594	23.7	4 425	14.3	2 543	8.9
Long products	2 401	13.7	7 182	22.4	6 398	20.7	8 059	28.2
Flat products	9 959	57.0	12 980	40.4	13 503	43.7	10 856	37.9
Wire	474	2.7	511	1.6	367	1.2	254	0.9
Tubes and fittings	2 647	15.2	3 836	11.9	6 182	20.0	254	0.9
Total	17 471	100.0	32 103	100.0	30 876	100.0	28 606	100.0

I. Far East

Years	1970		1974		1978		1982	
Products	1000t	%	1000t	%	1000t	%	1000t	%
Ingots and semis	49	7.6	126	9.6	131	6.6	1 327	26.9
Long products	479	73.9	183	14.0	500	25.2	872	17.7
Flat products	42	6.5	665	50.8	685	34.6	1 911	38.8
Wire	2	0.3	12	0.9	25	1.3	11	0.2
Tubes and fittings	76	11.7	324	24.7	641	32.3	811	16.4
Total	992		1 507		2 877		6 783	

Table 2.25. Structural changes in the exports of steel products (continued)

Years / Products	J. North America							
	1970		1974		1978		1982	
	1000t	%	1000t	%	1000t	%	1000t	%
Ingots and semis	3 005	49.1	854	20.1	350	16.0	417	14.2
Long products	675	11.0	744	17.5	503	23.0	469	15.9
Flat products	2 108	34.4	1 852	43.6	892	40.8	1 664	56.5
Wire	22	0.4	27	0.6	25	1.1	18	0.6
Tubes and fittings	311	5.1	773	18.2	416	19.0	376	12.8
Total	6 120	100.0	4 251	100.0	2 186	100.0	2 944	100.0

Years / Products	K. Latin America							
	1970		1974		1978		1982	
	1000t	%	1000t	%	1000t	%	1000t	%
Ingots and semis	19	3.5	3	1.0	354	24.8	248	7.9
Long products	250	45.9	190	63.7	489	34.4	900	28.8
Flat products	226	41.4	12	4.0	338	23.8	1 607	51.4
Wire	4	0.7	3	0.9	9	0.6	10	0.3
Tubes and fittings	46	8.5	91	30.4	234	16.4	361	11.5
Total	545	100.0	299	100.0	1 424	100.0	3 126	100.0

Years / Products	L. Oceania							
	1970		1974		1978		1982	
	1000t	%	1000t	%	1000t	%	1000t	%
Ingots and semis	445	55.5	797	77.1	1 591	63.5	644	51.0
Long products	36	4.5	28	2.7	341	13.6	148	11.7
Flat products	259	32.4	171	16.6	518	20.7	433	34.3
Wire	24	3.0	7	0.7	12	0.5	9	0.7
Tubes and fittings	37	4.6	30	2.9	46	1.8	29	2.3
Total	801	100.0	1 033	100.0	2 507	100.0	1 262	100.0

Table 2.25. Structural changes in the exports of steel products (continued)

Years Products	M. Africa							
	1970		1974		1978		1982	
	1000t	%	1000t	%	1000t	%	1000t	%
Ingots and semis	122	54.6	158	46.0	183	14.1	215	20.0
Long products	79	35.2	111	32.2	579	44.8	402	37.4
Flat products	22	9.9	67	19.5	484	37.5	377	35.1
Wire	1	0.3	3	0.9	19	1.4	8	0.8
Tubes and fittings	0	0.1	5	1.4	27	2.1	74	6.8
Total	224	100.0	343	100.0	1 292	100.0	1 075	100.0

CHAPTER III.   ANALYSIS OF STRUCTURAL CHANGES IN THE IMPORTS OF STEEL
PRODUCTS OF VARIOUS COUNTRIES AND/OR REGIONS

3.1   Analysis of changes in the geographic orientation of total imports of steel products by country and/or region

International trade in steel reflects, in part, the need of countries to import either all forms of steel or particular products, owing to the lack of a domestic source of supply.   But the greater part of international trade represents competitive inroads into domestic markets.   This latter kind of trade develops when imported steel has a better price/quality relationship than the domestic product.   This can result from the following factors: by reason of his location, the exporter may be better able to supply a particular market than the domestic supplier located less advantageously; costs may be lower at the prevailing exchange rate; the exporter may be differentiating between home and export prices; and the exporter may be receiving State assistance.   On the other hand, trade policy measures, taken as much by importing as by exporting countries, can influence the development of steel product exchanges.

Like domestic demand for steel and its production, international steel trade is subject to short-term fluctuation.   The reasons for this are, of course, manifold.   First of all, it should be borne in mind that the international steel market, although considerable in volume, is rather heterogeneous; it consists of the sum of steel demand of widely varying types, emanating from economies at very different stages of growth, using steel for rather different purposes and often transmitting their own short-term fluctuations to the international market.   To the fluctuations arising on the demand side must be added those which originate on the supply side.

Table 3.1 summarizes the geographic pattern of world steel imports (including and excluding intra-trade) and shows the major importing regions and their relative positions as importers.   As can be seen, the trade inside the regions is at least as significant as shipments to other regions: it may even be of greater importance. This is particularly true for the European Economic Community and eastern Europe, including the USSR.   Intra-regional trade for those groups of countries is larger than total shipments to countries outside the regions.   The trend over a long period suggests that the growing volume of intra-European trade in both eastern and western Europe is partially a reflection of a certain increase in the degree of international specialization.

Intra-Latin-American trade has never been significant compared with imports received by the region from outside.   Indeed, one of the main purposes of creating steel industries in these countries was to save foreign exchange previously expended on imported steel.   The Latin American countries have begun to export steel products to other regions.

It may thus be concluded that, wherever there is steel production on any scale, intra-regional trade throughout the period from 1970 was almost as important as trade between regions.

Below, an examination is made of steel trade excluding intra-regional trade.

During the period under review, the Far East was one of the major
importing regions of the world. Its imports showed an increasing trend and
more than doubled between 1970 and 1982, increasing from 8.3 million to
17.4 million tonnes. In 1978 the region's imports were even larger,
accounting for 22 million tonnes. Considering the size of the share of
Far East imports in total world imports, it should be mentioned that in 1970
it amounted to 13.3 per cent, by 1978 it had grown to 21.6 per cent and by
1982 it had again dropped to 17.6 per cent. Nevertheless, in 1982 the
Far East occupied first place among the importing regions of the world. The
major part of the region's imports went to China, India and the Republic of
Korea. In 1982 these countries imported 41 per cent of the total imports of
the region.

North America was another main importing region between 1970 and 1982.
In 1970, the North American countries occupied the leading position among the
importing regions of the world; they imported 12.0 million tonnes of steel
products and their share in total world steel imports amounted to
19.2 per cent. After 1970, the volume of North American imports did not
undergo any significant changes. Imports had grown to 16.8 million tonnes
by 1978 and by 1982 had dropped to almost the 1970 level (12.7 million
tonnes). As for the North American share in total world steel imports, there
was a downward trend between 1970 and 1982 when it declined from 19.2 to
12.8 per cent. During the period under review, the United States was the
principal importer of the region. In 1982, its imports accounted for
93.3 per cent of the total imports of the region and 11.9 per cent of the
total world imports.

The Middle East was a significant net importer of steel products. During
the period, the volume of its imports grew from 3.1 million to 14.1 million
tonnes. In total world imports of steel products, the Middle East occupied
ninth place in 1970 with its share of 5.0 per cent. Between 1970 and 1982 its
share increased to 14.2 per cent and the region attained third position
amongst world steel importers. The main importing country of the region had
become Saudi Arabia, which had increased its imports from 0.2 million tonnes
in 1970 to 5.1 million tonnes in 1982 and whose share in total imports of the
region accounted for 36.6 per cent in 1982.

The import volume of EEC members underwent the following changes between
1970 and 1982. In 1970 it amounted to 8.5 million tonnes, by 1974 it had
dropped to 6.4 million tonnes and by 1982 it had again increased to 9 million
tonnes. As for the EEC share in total world steel imports, it recorded a
downward trend. Accounting for 13.6 per cent in 1970, by 1982 the share had
decreased to 9.1 per cent. The major importing country of the region was the
Federal Republic of Germany, which in 1970 imported 2.5 million tonnes of
steel products or 30 per cent of the total steel imports of the region. By
1982, the Federal Republic, importing 3.5 million tonnes, had increased its
share to 38.4 per cent. The other principal importers of the region were
Italy, the United Kingdom and France.

In 1970, the USSR imported 2.2 million tonnes of steel products and its
share in total world imports accounted for 3.6 per cent. According to
estimates, USSR steel imports had increased to 10.1 million tonnes by 1982 and
its share in total world imports to 10.2 per cent. In 1982 the USSR occupied
fourth place among the importing regions of the world.

The steel imports of the east European countries amounted to 6.9 million tonnes in 1970. During the period from 1970 to 1978 they increased to 8.6 million tonnes, at the same time as the share of the region's imports in total world imports dropped from 11.1 to 8.3 per cent. After 1978, there was a downward trend both in the volume of imports and in the share, the volume declining to 7.2 million tonnes and the share to 7.3 per cent. The German Democratic Republic was the main steel importer of the region between 1970 and 1982, importing 2.8 million tonnes or 39.6 per cent of the total imports of the region in 1970. By 1982 its imports had increased to 3.5 million tonnes, accounting for 48.3 per cent of the region's imports. Romania occupied second position among the importing countries of the region. The volume of its imports was relatively stable - at around 1.1 million tonnes - between 1970 and 1978, but by 1982 it had dropped to 0.5 million tonnes and the country's share in the region's imports had declined from 13.1 to 7.0 per cent. It must be pointed out that, owing to the lack of proper data concerning the distribution of imports of steel products within the east European countries, a part of the imports of this region was included under "eastern Europe unallocated". The data concerning the USSR exports to this region for the period from 1977 to 1982 were estimated using the methodology described in chapter I.

Between 1970 and 1974, the Latin American countries considerably increased the volume of their steel imports, from 4.4 million to 10.2 million tonnes; during the same period, their share in total world imports grew from 7.0 to 11.0 per cent. But after 1974 the situation changed significantly and a sharp decrease was recorded both in the volume of the region's imports and in their share, which fell to 5.0 million tonnes and 5.1 per cent, respectively, by 1982. This was mainly due to the fact that Brazil, which was the major importing country of the region, importing 4.1 million tonnes of steel products in 1974, had strongly decreased the volume of its imports, which by 1982 had dropped to 0.4 million tonnes.

The volume of steel imports of the North European countries showed only slight modifications between 1970 and 1982. The region's imports grew from 3.3 million tonnes in 1970 to 4.1 million tonnes in 1974. Following a decrease in 1978 (2.7 million tonnes), they increased again up to 1982, attaining 3.8 million tonnes in 1982. The share of the region in total world steel imports also fluctuated between 1970 and 1982, at the same time showing a declining trend; it accounted for 5.2 per cent in 1970 and dropped to 3.8 per cent in 1982. The main importing countries of the region were Norway and Sweden, which together imported 86.6 per cent of the total steel imports of the region.

In 1970, the south European countries, importing 5.0 million tonnes of steel products, occupied fifth place in the world among importing regions and their share in total world steel imports accounted for 8.0 per cent. Between 1970 and 1975, the region's imports grew to 6.6 million tonnes. During the next period, up to 1982, south European imports were rather stable, accounting for around 5.0 million tonnes. Their share in total world steel imports had undergone the following changes: in 1974 it accounted for 7.0 per cent (a decrease compared with 1970), then it dropped again in 1978 (5.3 per cent) and grew to 5.8 per cent in 1982. The main importing countries of the region were Greece, Spain and Yugoslavia. In 1982 these three countries accounted for 76.8 per cent of the total steel imports of the region.

As can be seen from table 3.1, the volume of the central European countries' steel imports did not undergo any significant changes during the period. Amounting to 2.3 million tonnes in 1970, it increased slightly in 1978 (2.7 million tonnes) and, in 1982, reached almost the same level (2.4 million tonnes) as in 1970. The share of the region in total world steel imports was also relatively constant. Between 1970 and 1973 it accounted for around 3.8 per cent and during the following period, from 1974 to 1982, it fluctuated between 2.5 and 2.7 per cent. Switzerland was the principal importer of steel products in the region, accounting for around 70 per cent of the region's imports.

African steel imports grew from 3.3 million tonnes in 1970 to 5.5 million tonnes in 1974. Following a decrease in 1978 (5.0 million tonnes), they grew again up to 1981, when they reached their peak (6.3 million tonnes) and declined to 5.1 million tonnes in 1982. The share of African imports in total world steel imports changed in the following way: in 1970 it accounted for 5.3 per cent, increasing to 5.9 per cent in 1974 and dropping to 5 per cent by 1978; it grew again in 1981 (6.3 per cent) and once more dropped to 5.1 per cent in 1982. Between 1970 and 1982, Algeria was the major importing country of the region; its share in total imports of the region increased from 19.1 per cent in 1970 to 27.9 per cent in 1982.

Oceania imported a small proportion of the total world imports of steel products. Its share amounted to only 1.4 per cent in 1970, reaching its maximum (1.7 per cent) in 1974 and, after that, fluctuating at around 1.0 per cent until, by 1982, it accounted for 1.2 per cent. During the same period the volume of the region's steel imports grew from 0.9 million tonnes in 1970 to 1.6 million tonnes in 1974; it dropped to 0.8 million tonnes by 1978 and grew again in 1982 (1.2 million tonnes).

Between 1970 and 1976 the volume of Japanese steel imports was very small. Imports amounted to 0.1 million tonnes in 1970 and, with some fluctuations, grew to 0.3 million tonnes by 1974. During the following two years, Japan imported less than 0.1 million tonnes of steel products. But after 1976 the situation changed significantly. Between 1976 and 1982 there was a strong upward trend, when imports increased from 0.09 million to 1.94 million tonnes. The share of Japanese imports in total world steel imports underwent the same evolution between 1970 and 1982. Accounting for 0.2 per cent in 1970, it increased to 0.3 per cent in 1974. Following a decline in 1976 (0.1 per cent), it grew again up to 1982, when the share reached its maximum - 2.0 per cent. Thus in 1982 Japan was no longer in last position amongst the importing regions of the world, although its imports remained low when compared with its national consumption or with the imports of the other main world steel producers.

Table 3.1. Changes in the geographic orientation of imports of steel products

| | World (including intra-trade) | | | | | | | | World (excluding intra-trade) | | | | | | | |
| Years | 1970 | | 1974 | | 1978 | | 1982 | | 1970 | | 1974 | | 1978 | | 1982 | |
Importers	1000t	%	1000t	%	1000t	%	1000t	%	1000t	%	1000t	%	1000t	%	1000t	%
Africa	3389	3.8	5525	4.3	5124	3.8	5114	3.9	3332	5.3	5461	5.9	5053	5.0	5103	5.1
Algeria	634	0.7	1135	0.9	1478	1.1	1423	1.1	634	1.0	1135	1.2	1478	1.4	1423	1.4
Liby.Arab Jam.	198	0.2	619	0.5	419	0.3	302	0.2	198	0.3	619	0.7	419	0.4	302	0.3
Morocco	252	0.3	349	0.3	483	0.4	602	0.5	252	0.4	349	0.4	483	0.5	602	0.6
Tunisia	65	0.1	113	0.1	184	0.1	348	0.3	65	0.1	113	0.1	184	0.2	348	0.4
North Africa	1150	1.3	2216	1.7	2565	1.9	2676	2.0	1150	1.8	2216	2.4	2565	2.5	2676	2.7
South Africa	439	0.5	1115	0.9	137	0.1	167	0.1	439	0.7	1115	1.2	137	0.1	167	0.2
Other Africa	1800	2.0	2194	1.7	2422	1.8	2270	1.7	1743	2.8	2130	2.3	2352	2.3	2260	2.3
Far East	8473	9.4	14425	11.2	22673	16.7	20615	15.8	8342	13.4	14249	15.4	22125	21.7	19085	19.3
China	2096	2.3	3579	2.8	8908	6.6	3766	2.9	2096	3.4	3579	3.9	8894	8.7	3765	3.8
India	726	0.8	1309	1.0	1099	0.8	2478	1.9	726	1.2	1306	1.4	1088	1.1	2112	2.1
Korea,Rep.of	483	0.5	1902	1.5	2971	2.2	1254	1.0	475	0.8	1902	2.1	2956	2.9	1254	1.3
3 of Far East	3306	3.7	6790	5.3	12978	9.5	7499	5.7	3298	5.3	6788	7.3	12938	12.7	7132	7.2
Japan	117	0.1	301	0.2	371	0.3	371	1.5	117	0.2	301	0.3	371	0.4	1941	2.0
Other Far East	5051	5.6	7334	5.7	9323	6.9	11175	8.6	4926	7.9	7160	7.7	8816	8.6	10012	10.1
Middle East	3116	3.5	7646	5.9	10568	7.8	14061	10.8	3116	5.0	7646	8.3	10568	10.4	14061	14.2
Egypt	378	0.4	490	0.4	486	0.4	999	0.8	378	0.6	490	0.5	486	0.5	999	1.0
Saudi Arabia	194	0.2	738	0.6	2070	1.5	5143	3.9	194	0.3	738	0.8	2070	2.0	5143	5.2
2 of Middle East	572	0.6	1227	0.9	2556	1.9	6142	4.7	572	0.9	1227	1.3	2556	2.5	6142	6.2
Other Middle East	2543	2.8	6418	5.0	8012	5.9	7919	6.1	2543	4.1	6418	6.9	8012	7.8	7919	8.0
Oceania	1140	1.3	1826	1.4	1026	0.8	1464	1.1	885	1.4	1586	1.7	822	0.8	1221	1.2
Australia	480	0.5	916	0.7	480	0.4	794	0.6	480	0.8	915	1.0	477	0.5	780	0.8
New Zealand	534	0.6	790	0.6	455	0.3	560	0.4	337	0.5	607	0.7	307	0.3	402	0.4
Other Oceania	126	0.1	120	0.1	91	0.1	111	0.1	67	0.1	63	0.1	39	0.0	39	0.0
North America	13648	15.2	17753	13.7	19746	14.5	14764	11.3	12009	19.2	15134	16.4	16841	16.5	12688	12.6
Canada	1488	1.7	3243	2.5	1786	1.3	1154	0.9	798	1.3	1805	2.0	1074	1.1	826	0.8
United States	12153	13.5	14503	11.2	17950	13.2	13589	10.4	11206	17.9	13323	14.4	15758	15.4	11843	11.9
N. America Unalloc.	6	0.0	8	0.0	10	0.0	21	0.0	5	0.0	6	0.0	9	0.0	19	0.0
Other America	4987	5.5	10648	8.2	7653	5.6	5653	4.3	4365	7.0	10200	11.0	7064	6.9	5008	5.1
Argentina	1571	1.7	1646	1.3	591	0.4	636	0.5	1229	2.0	1531	1.7	562	0.6	484	0.5
Brazil	549	0.6	4234	3.3	663	0.5	407	0.3	525	0.8	4083	4.4	589	0.6	362	0.4
Mexico	268	0.3	738	0.6	1841	1.4	1127	0.9	222	0.4	737	0.8	1833	1.8	1104	1.1
Venezuela	518	0.6	1316	1.0	1737	1.3	970	0.7	505	0.8	1308	1.4	1565	1.6	880	0.9
4 of Other America	2905	3.2	7934	6.1	4831	3.6	3141	2.4	2481	4.0	7658	8.3	4569	4.5	2829	2.9
Other Other America	2082	2.3	2713	2.1	2822	2.1	2512	1.9	1884	3.0	2542	2.8	2515	2.5	2180	2.2
Western Europe	42763	47.5	51068	39.5	47940	35.3	46672	35.7	19206	30.8	19606	21.2	20105	19.7	21133	21.3
B.L.E.U.	2379	2.6	3705	2.9	3129	2.3	3204	2.5	562	0.9	603	0.7	826	0.8	803	0.8
Denmark	1438	1.6	1585	1.2	1279	0.9	1676	1.3	484	0.8	468	0.5	417	0.4	549	0.6
France	7505	8.3	8198	6.3	8306	6.1	7633	5.8	996	1.6	536	0.6	937	0.9	1088	1.1
Germany,Fed.Rep.	9101	10.1	9272	7.2	11097	8.2	9031	6.9	2546	4.1	2232	2.4	3639	3.6	3473	3.5
Ireland	230	0.3	421	0.3	392	0.3	394	0.3	17	0.0	19	0.0	53	0.1	53	0.1
Italy	4787	5.3	4658	3.6	4467	3.3	4622	3.7	2008	3.2	1188	1.3	1313	1.3	1414	1.4
Netherlands	3562	4.0	4931	3.8	3604	2.7	3006	2.3	272	0.4	349	0.4	531	0.5	478	0.5
United Kingdom	2224	2.5	3782	2.9	3506	2.6	3723	2.9	1595	2.6	1026	1.1	1314	1.3	1182	1.2
EEC(9)	31226	34.7	36552	28.3	35781	26.3	33488	25.6	8481	13.6	6421	6.9	9031	8.8	9040	9.1
Finland	856	1.0	801	0.6	424	0.3	627	0.5	716	1.1	603	0.7	291	0.3	483	0.5
Iceland	36	0.0	49	0.0	41	0.0	43	0.0	30	0.0	35	0.0	25	0.0	25	0.0
Norway	1196	1.3	1697	1.3	1131	0.8	1860	1.4	1015	1.6	1464	1.6	893	0.9	1590	1.6
Sweden	1678	1.9	2337	1.8	1839	1.4	1939	1.5	1502	2.4	2000	2.2	1492	1.5	1682	1.7
Northern Europe	3766	4.2	4884	3.8	3436	2.5	4469	3.4	3264	5.2	4102	4.4	2700	2.6	3781	3.8
Greece	958	1.1	1141	0.9	1284	0.9	1071	0.8	940	1.5	1118	1.2	1225	1.2	1057	1.1
Portugal	475	0.5	928	0.7	675	0.5	793	0.6	463	0.7	912	1.0	621	0.6	700	0.7
Spain	1860	2.1	1461	1.1	808	0.6	1896	1.5	1858	3.0	1348	1.5	807	0.8	1894	1.9
Turkey	323	0.4	1366	1.1	685	0.5	696	0.5	321	0.5	1291	1.4	645	0.6	629	0.6
Yugoslavia	1538	1.7	1882	1.5	2117	1.6	1524	1.2	1431	2.3	1787	1.9	2084	2.0	1454	1.5
Southern Europe	5154	5.7	6798	5.3	5569	4.1	5980	4.6	5013	8.0	6457	7.0	5382	5.3	5734	5.8
Austria	482	0.5	707	0.5	940	0.7	872	0.7	466	0.7	650	0.7	897	0.9	826	0.8
Switzerland	2032	2.3	1910	1.5	1928	1.4	1726	1.3	1880	3.0	1760	1.9	1809	1.8	1614	1.6
Central Europe	2514	2.8	2617	2.0	2868	2.1	2597	2.0	2346	3.8	2410	2.6	2706	2.7	2440	2.5
W.Europe unalloc	102	0.1	217	0.2	286	0.2	136	0.1	102	0.2	217	0.2	286	0.3	136	0.1
Eastern Europe	10425	11.6	17970	13.9	19253	14.2	18641	14.3	9191	14.7	16199	17.5	17461	17.1	17227	17.4
Albania	55	0.1	97	0.1	89	0.1	124	0.1	20	0.0	44	0.0	29	0.0	51	0.1
Bulgaria	976	1.1	1053	0.8	483	0.4	409	0.3	895	1.4	933	1.0	327	0.3	231	0.2
Czechoslovakia	582	0.6	365	0.3	326	0.2	181	0.1	533	0.9	203	0.2	194	0.2	57	0.1
German Dem.Rep.	3048	3.4	3324	2.6	3656	2.7	3914	3.0	2750	4.4	2917	3.2	3120	3.1	3456	3.5
Hungary	832	0.9	999	0.8	339	0.2	188	0.1	693	1.1	760	0.8	176	0.2	106	0.1
Poland	1246	1.4	3103	2.4	1065	0.8	399	0.3	960	1.5	2622	2.8	783	0.8	115	0.1
Romania	1435	1.6	1360	1.1	1569	1.2	716	0.5	1090	1.7	1051	1.1	1125	1.1	502	0.5
7 of E.Europe	8175	9.1	10301	8.0	7527	5.5	5930	4.5	6941	11.1	8529	9.2	5756	5.6	4516	4.6
USSR	2250	2.5	7619	5.9	8924	6.6	10072	7.7	2250	3.6	7619	8.2	8924	8.7	10072	10.2
E.Europe unalloc	0	0.0	50	0.0	2802	2.1	2639	2.0	0	0.0	50	0.1	2802	2.7	2639	2.7
Unallocated	2006	2.2	2354	1.8	1988	1.5	3585	2.7	2006	3.2	2354	2.5	1988	1.9	3585	3.6
World	89948	100.0	129215	100.0	135971	100.0	130567	100.0	62450	100.0	92436	100.0	102069	100.0	99111	100.0

3.2 Evaluation of the relative importance of each importer and/or importing region for various exporting countries and/or regions

Africa

The changes which occurred between 1970 and 1982 in the importance of Africa as a destination of steel exports for various exporting regions are presented in table 3.2.

It will be seen that, in 1970, Africa imported 23.9 per cent of the total steel exports of the Latin American countries. Between 1970 and 1982, however, the situation changed significantly, with Africa losing its importance as a destination of steel exports from those countries. By 1982 the share of Latin American steel exports going to Africa accounted for only 6.2 per cent.

In 1970 Africa was importing 10.2 per cent of total EEC steel exports. This share remained stable until 1974 at 10.5 per cent and then decreased slightly, fluctuating around 9 per cent.

During the last years of the period under review, Africa grew in importance as a destination of steel exports from the southern countries of western Europe. The share of those exports increased from 7.4 per cent in 1970 to 19.8 per cent in 1982.

As far as African steel imports from other main exporting regions are concerned, Africa imported a rather small proportion of their total steel exports. In this context, only two exporting regions need be mentioned - Japan and North America. In 1980, Africa imported, respectively, 3.3 per cent and 5.0 per cent of their total steel exports.

Oceania

Between 1970 and 1982, Oceania was of interest as an importer of steel products from Japan and the Far East (see table 3.2). In 1970, Oceania imported 3.4 per cent of the total Japanese steel exports and, by 1974, the share had increased to 3.9 per cent. Following a decline in 1978 (2 per cent), it grew again, attaining 3.3 per cent by 1982. Oceanian steel imports from the Far East reached their peak in 1974, at which time they accounted for 4.6 per cent of the total steel exports of the Far East. This was followed by a downward trend in the importance of Oceania as a destination of Far Eastern steel exports.

All other exporting regions delivered only a very small proportion of their steel exports to Oceania during the period. The east European countries and the USSR did not export any steel to the region.

Far East

The Far East was of great interest as a recipient of steel exports from the following regions: Japan, Oceania, North America and Latin America.

Table 3.2.   Imports of steel products: Africa and Oceania

Africa

Exporters	1970 1000t	1970 %	1974 1000t	1974 %	1978 1000t	1978 %	1982 1000t	1982 %
Africa	.		.		.		80	1.2
Far East	35	3.5	29	1.9	58	2.0		
Japan	792	4.5	1353	4.2	1054	3.4	931	3.3
Oceania	11	1.4	11	1.0	13	0.5	2	0.2
North America	204	3.3	293	6.9	101	4.6	148	5.0
Latin America	130	23.9	14	4.8	94	6.6	195	6.2
EEC (9)	1929	10.2	3466	10.5	2989	9.0	2379	9.2
Northern Europe	15	0.8	48	2.1	78	2.4	51	1.7
Southern Europe	46	7.4	106	6.9	498	10.5	1214	19.8
Central Europe	9	0.7	15	0.9	39	1.6	33	1.2
Eastern Europe	106	1.7	72	1.0	102	1.0	64	0.7
USSR	55	0.7	54	0.8	26	0.4	6	0.1
World	3332	5.3	5461	5.9	5053	5.0	5103	5.1

Oceania

Exporters	1970 1000t	1970 %	1974 1000t	1974 %	1978 1000t	1978 %	1982 1000t	1982 %
Africa	5	2.0	2	0.4	1	0.1	0	0.0
Far East	8	0.8	70	4.6	76	2.7	103	1.5
Japan	594	3.4	1248	3.9	618	2.0	940	3.3
Oceania								
North America	55	0.9	67	1.6	17	0.8	15	0.5
Latin America	0	0.0	0	0.0	1	0.1	22	0.7
EEC (9)	210	1.1	182	0.6	97	0.3	129	0.5
Northern Europe	10	0.5	12	0.5	7	0.2	7	0.2
Southern Europe	1	0.1	0	0.0	3	0.1	1	0.0
Central Europe	3	0.2	4	0.2	2	0.1	3	0.1
Eastern Europe	0	0.0	1	0.0	-	-	0	0.0
USSR	-	-		-		-		-
World	885	1.4	1586	1.7	822	0.8	1221	1.2

Table 3.3.   Imports of steel products: Far East and Japan

Far East

Exporters	1970 1000t	1970 %	1974 1000t	1974 %	1978 1000t	1978 %	1982 1000t	1982 %
Africa	5	2.2	9	2.5	286	22.2	88	8.2
Far East	.		.		.		.	
Japan	5252	30.1	10829	33.7	14055	45.5	12000	41.9
Oceania	472	58.9	632	61.2	1517	60.5	848	67.2
North America	836	13.7	605	14.2	555	25.4	699	23.8
Latin America	0	0.0	0	0.2	204	14.3	627	20.0
EEC (9)	1215	6.4	1592	4.8	4287	12.9	2174	8.4
Northern Europe	60	3.2	41	1.8	174	5.3	103	3.3
Southern Europe	24	3.8	50	3.3	410	8.7	201	3.3
Central Europe	10	0.8	11	0.6	24	1.0	12	0.4
Eastern Europe	213	3.5	252	3.3	420	4.3	384	3.9
USSR	191	2.6	204	5.0	95	1.3	255	1.3
World	8278	13.3	14225	15.4	22025	21.6	17391	17.5

Japan

Exporters	1970 1000t	1970 %	1974 1000t	1974 %	1978 1000t	1978 %	1982 1000t	1982 %
Africa	0	0.0	0	0.0	0	0.0	0	0.0
Far East	64	6.4	24	1.6	99	3.5	1694	25.0
Japan			-	-	-	-	-	-
Oceania	11	1.4	247	24.0	65	2.6	.	0.0
North America	16	0.3	12	0.3	54	2.4	14	0.5
Latin America	0	0.0	0	0.0	26	1.8	221	7.1
EEC (9)	5	0.0	11	0.0	10	0.0	8	0.0
Northern Europe	4	0.2	5	0.2	27	0.8	3	0.1
Southern Europe	0	-	0	-	34	0.7	0	0.0
Central Europe	-	0.0	1	0.0	0	0.0	0	0.0
Eastern Europe	16	0.3	1	0.0	56	0.6	-	-
USSR	-	-	-	-	-	-	-	-
World	117	0.2	301	0.3	371	0.4	1941	2.0

As can be seen from table 3.3, between 1970 and 1982, the Far East imported a considerable proportion of Japanese steel products; in 1970, this amounted to 30.1 per cent and, by 1978, it had increased to 45.5 per cent, almost half of total Japanese steel exports. In 1982, the share showed a decline of around 4 per cent.

The Far East was the major destination of steel exports from Oceania. Between 1970 and 1982, the region imported the major part of the Oceanian steel products, the share growing from 58.9 to 67.2 per cent during the period.

A considerable proportion of North American steel exports went to the Far East. Its share, accounting for 13.7 per cent in 1970, increased to 25.4 per cent by 1978 and then dropped slightly in 1982, to 23.8 per cent.

In the second part of the 1970s, the Far East began to import steel products from the Latin American countries and gained in importance as a destination of steel exports from these countries. Between 1974 and 1982, the share of the Far East in Latin American steel exports increased from 0.2 to 20.0 per cent.

The share of EEC steel products imported by the Far East fluctuated between 5 and 13 per cent during the period.

Four regions - the northern and southern countries of western Europe, the east European countries and the USSR - exported almost identical proportions of their total steel exports to the region. The shares were not constant year by year but, in general, amounted to around 3 per cent.

As for the importance of the Far East as an importer of African steel, it will be noted that, while some years the region imported around one fifth of the total exports, for the main part of the period the share was significantly lower, accounting, for example, for only 2.2 per cent in 1970 and 8.2 per cent in 1982.

Japan

As from 1979, the major part of Japanese steel imports came from the Far East. While in 1974, the share of the Far East steel exports delivered to Japan accounted for only 1.6 per cent, by 1982 it had increased to 25.0 per cent, making Japan the principal steel importer of the Far East (see table 3.3).

Another very interesting development which occurred between 1970 and 1982 was the increasing interest shown by the Latin American countries in the Japanese steel market. The share of the Latin American steel exports to Japan increased from practically 0 to 7.1 per cent. Japanese imports from other main steel exporting regions were insignificant and the shares in total steel deliveries of the regions relatively small.

European Economic Community

In 1970, EEC was the main recipient of steel exports from the following regions (see table 3.4):

-       Northern countries of western Europe, with an export share of 77.3 per cent;

-       Central countries of western Europe, with an export share of 64.8 per cent;

-       North America, with an export share of 48.3 per cent; and

-       Southern countries of western Europe, with an export share of 41.1 per cent.

About one fourth of the total steel exports of Africa and the east European countries were imported by EEC.

The main changes in the relative importance of EEC for various exporting regions between 1970 and 1982 were:

-       Considerable decrease in the export shares of the North American countries and the southern countries of western Europe directed to the region. By 1982 the shares had dropped to 14.6 and to 18.9 per cent, respectively;

-       After importing 24.3 per cent of the steel exports from Oceania in 1974, EEC decreased this share to 2.8 per cent by 1982;

-       There was a downward trend in the share of Japanese steel exports, which declined from 5.9 to 0.9 per cent; and

-       The share of the Latin American countries' steel exports to EEC, which declined from 14.7 per cent in 1970 to 5.2 per cent in 1974, grew to 18.8 per cent by 1982.

Northern countries of western Europe

Modifications in the relative importance of the northern countries of western Europe as an importer of steel products for various exporting regions are presented in table 3.4. As can be seen, between 1970 and 1982, there were no significant changes in the shares of the region's steel exports to the north European countries.

While EEC continued sending the larger proportion of its steel exports to these countries, during the period, the share dropped from 13.3 to 8.5 per cent.

In 1982, the north European countries imported 8.7 per cent of the steel exports of the Far East; between 1970 and 1981, the region had imported almost no steel products from the Far East.

The export shares of all other exporting regions, with the exception of that of the south European countries, had dropped slightly by 1982.

Table 3.4. Imports of steel products: European Economic Community and the northern countries of western Europe

Exporters	European Economic Community								Northern countries of western Europe							
	1970		1974		1978		1982		1970		1974		1978		1982	
	1000t	%	1000t	%	1000t	%	1000t	%	1000t	%	1000t	%	1000t	%	1000t	%
Africa	61	27.2	132	38.5	173	13.4	303	28.2	0	0.2	1	0.2	1	0.2	0	0.0
Far East	23	2.3	38	2.5	86	3.0	145	2.1	0	0.0	0	0.0	7	0.0	589	8.7
Japan	1038	5.9	1054	5.9	623	2.0	260	0.9	201	1.2	394	1.2	118	1.2	262	0.9
Oceania	97	12.1	251	24.3	346	13.8	35	2.8	24	3.0	0	0.0	0	0.0	0	0.0
North America	2956	48.3	497	11.7	264	12.1	430	14.6	90	1.5	39	0.9	7	0.3	16	0.5
Latin America	80	14.7	16	5.2	152	10.6	589	18.8	10	1.8	1	0.2	0	0.0	22	0.7
EEC (9)									2521	13.3	3230	9.8	1980	5.9	2205	8.5
Northern Europe	1421	77.3	1567	68.7	2085	63.6	1994	64.9	.							
Southern Europe	255	41.1	434	28.1	1177	24.9	1156	18.9	9	1.4	13	0.8	181	3.8	195	3.2
Central Europe	844	64.8	1005	57.5	1626	66.0	1764	61.0	65	5.0	112	6.4	105	4.3	108	3.7
Eastern Europe	1571	25.6	1415	18.7	2356	24.2	2220	22.6	248	4.0	260	3.4	288	3.0	352	3.6
USSR	132	1.8	12	0.2	143	1.9	145	1.9	96	1.3	53	0.8	12	0.8	32	0.4
World	8481	13.6	6421	6.9	9031	8.8	9040	9.1	3264	5.2	4102	4.4	2700	2.6	3781	3.8

Table 3.5. Imports of steel products: Southern and central countries of western Europe

Exporters	Southern countries of western Europe								Central countries of western Europe							
	1970		1974		1978		1982		1970		1974		1978		1982	
	1000t	%	1000t	%	1000t	%	1000t	%	1000t	%	1000t	%	1000t	%	1000t	%
Africa	46	20.5	15	4.3	51	4.3	105	9.8	1	0.5	0	0.1	0	0.0	6	0.5
Far East	11	1.1	20	1.4	1	0.1	24	0.3	-	-	0	0.0	1	0.0	2	0.0
Japan	1027	5.9	1333	4.2	516	1.7	312	1.1	11	0.1	63	0.2	85	0.3	19	0.1
Oceania	38	4.7	0	0.0	0	0.0	10	0.8	-	-	0	0.0	-	-	-	-
North America	563	9.5	228	5.4	39	1.8	226	7.7	75	1.2	6	0.1	8	0.4	21	0.7
Latin America	39	7.2	4	1.5	171	12.0	78	2.5		0.0	-	-	-	-	13	0.4
EEC (9)	2006	10.6	3380	10.3	2447	7.3	3352	13.0	1907	10.0	2019	6.1	1966	5.9	1864	7.2
Northern Europe	75	4.1	111	4.9	42	1.3	167	5.4	42	2.3	39	1.7	54	1.7	53	1.7
Southern Europe									11	1.7	21	1.4	153	3.2	151	2.5
Central Europe	55	4.2	127	7.3	185	7.5	138	4.8								
Eastern Europe	827	13.5	866	11.5	1657	17.0	1025	10.4	299	4.9	260	3.4	438	4.5	311	3.2
USSR	306	4.1	373	5.5	273	3.7	296	3.9	1	0.0	2	0.0	1	0.0	1	0.0
World	5013	8.0	6457	7.0	5382	5.3	5734	5.8	2346	3.8	2410	2.6	2706	2.7	2440	2.5

Southern countries of western Europe

The share of EEC steel products imported by the southern countries of western Europe was around 10 per cent per year and, in 1982, it grew to 13 per cent (see table 3.5). When considering the importance of these countries as an import market for other exporting regions, the east European countries should be mentioned, since the share of their exports to the south European countries amounted to 10.4 per cent in 1982.

In 1970, the region imported 20.5 per cent of the total steel exports of Africa. Since then, however, following considerable fluctuation, the share dropped to 9.8 per cent in 1982. A decline was recorded by the countries of North America in the share of their exports imported by the region. Accounting for 9.5 per cent in 1970, it dropped to 1.8 per cent in 1978 and, following an increase, again attained 7.7 per cent in 1982. The export shares of the North European and the central European countries of western Europe and the USSR did not change significantly between 1970 and 1982.

A general downward trend was recorded in the shares of exports of Japan and the Latin American countries to the southern countries of western Europe. Between 1970 and 1982, they dropped, with some fluctuation, from 5.9 to 1.1 per cent and from 7.2 to 2.5 per cent, respectively. Other exporting regions, with previously small shares, saw their shares drop even lower by 1982.

Central countries of western Europe

It may be seen in table 3.5 that only the EEC and the east European countries sent a sizeable proportion of their steel exports to the central countries of western Europe. However, it should be noted that, between 1970 and 1982, these shares dropped from 10.0 to 7.2 per cent (EEC) and from 4.9 to 3.2 per cent (east European countries).

Other steel-exporting regions directed only a small proportion of their exports to the region, their shares remaining virtually stable.

East European countries

The changes which occurred between 1970 and 1982 in the importance of the east European countries as an importing region for various exporting regions are shown in table 3.6. As can be seen, the region formed a principal market for USSR steel exports during the period, with more than 80 per cent going to the region.

Other exporting regions having an interest in the region's steel market were the following:

- EEC. Its share, accounting for 4.1 per cent in 1970, increased to 7.4 per cent in 1974 (in 1976 it was even higher - 10.0 per cent) and, by 1982, dropped to 1.7 per cent;

- The southern countries of western Europe. Their share dropped from 19.3 to 3.7 per cent;

- The central countries of western Europe. Their share, while fluctuating, dropped from 10.6 to 2.6 per cent; and

- The northern countries of western Europe. Their share underwent considerable fluctuation between 1970 and 1982. Amounting to 2.4 per cent in 1970, it increased to 6.9 per cent by 1974 and dropped to 1.3 per cent in 1982.

USSR

As indicated in table 3.6, there were four exporting regions which sent a relatively significant proportion of their steel exports to the USSR between 1970 and 1982. They are: the east European countries, EEC, the central countries of western Europe and Japan.

The share of steel exports from the east European countries imported by the USSR showed an upward trend between 1970 and 1982, increasing from 11.7 to 26.3 per cent. In other words, in 1982 the USSR imported more than one fourth of the total exports of those countries.

A similar development was noted of the share of EEC and the central countries of western Europe, which grew from 5 to 16.2 per cent and from 7.5 to 18.3 per cent, respectively, between 1970 and 1982.

During the period, the share of Japanese exports imported by the USSR increased considerably. It grew from 1.2 per cent in 1970 to 8.3 per cent in 1982.

After recording a share of 13.8 per cent in 1970, the Far East had practically stopped exporting steel products to the USSR by 1974. After that the share fluctuated and, by 1982, accounted for only 0.2 per cent of the total steel exports of the Far East.

The shares of steel exports of the northern and the southern countries of western Europe imported by the USSR showed little change between 1970 and 1982. Accounting for 2.3 per cent and 5.2 per cent in 1970, they amounted to 2.4 per cent and 4.4 per cent, respectively, in 1982.

North America

Between 1970 and 1982, the North American countries were the major destination for Latin American steel exports (see table 3.7). In 1970, their share accounted for 50.8 per cent. This was followed by a large increase in 1971, when the share reached its maximum of 93.2 per cent, and a period, between 1971 and 1978, when it varied at around 85 per cent. After 1978 and up to 1982, there was a downward trend in the share of steel exports from the Latin American countries to the region, with the share declining from 86.1 per cent in 1978 to 32.7 per cent in 1982.

Between 1970 and 1982 the region imported a considerable proportion of Japanese steel exports. The main feature of the changes which occurred in the relative importance of the region for Japan was the decrease in the share of Japanese steel exports to the region, which dropped from 33.8 per cent in 1970 to 14.6 per cent in 1982.

Table 3.6. Imports of steel products: East European countries and the USSR

Exporters	East European countries								USSR							
	1970		1974		1978		1982		1970		1974		1978		1982	
	1000t	%	1000t	%	1000t	%	1000t	%	1000t	%	1000t	%	1000t	%	1000t	%
Africa	2	0.2	2	0.2	-	-	-	-	-	-	0	0.0	-	-	-	-
Far East	248	1.4	259	0.8	148	0.5	41	0.1	137	13.8	-	-	86	3.0	15	0.2
Japan	28	3.5	20	1.9	0	0.0	-	-	218	1.2	1317	4.1	1516	4.9	2380	8.3
Oceania	40	0.6	41	1.0	32	1.5	-	-	-	-	24	0.6	8	0.4	-	-
North America	0	0.0	0	0.0	2	0.2	1	0.0	48	0.8	1	0.2	13	0.9	2	0.1
Latin America	-	-	-	-	-	-	0	0.0	-	-	-	-	-	-	16	0.5
EEC (9)	782	4.1	2421	7.4	1574	4.7	438	1.7	958	5.0	4149	12.6	4466	13.4	4196	16.2
Northern Europe	44	2.4	158	6.9	87	2.7	41	1.3	43	2.3	34	1.5	106	3.3	75	2.4
Southern Europe	120	19.3	281	18.2	235	5.0	228	3.7	32	5.2	228	14.7	437	9.2	272	4.4
Central Europe	138	10.6	197	11.2	143	5.8	74	2.6	98	7.5	153	8.8	262	10.6	529	18.3
Eastern Europe	-	-	-	-	-	-	-	-	715	11.7	1713	22.7	2029	20.8	2586	26.3
USSR	5540	74.8	5200	76.0	6336	86.0	6331	83.6	-	-	-	-	-	-	-	-
World	6941	11.1	8579	9.3	8557	8.4	7154	7.2	2250	3.6	7619	8.2	8924	8.7	10072	10.2

Table 3.7. Imports of steel products: North America and Latin America

Exporters	North America								Latin America							
	1970		1974		1978		1982		1970		1974		1978		1982	
	1000t	%	1000t	%	1000t	%	1000t	%	1000t	%	1000t	%	1000t	%	1000t	%
Africa	58	26.1	20	5.9	644	49.9	510	47.4	22	10.0	33	9.7	99	7.7	64	5.9
Far East	87	8.8	858	57.0	1058	36.8	924	13.6	2	0.2	118	7.9	7	0.2	102	1.5
Japan	5907	33.8	6675	33.8	5996	19.4	4170	14.6	1566	9.0	4350	13.6	2415	7.8	1478	5.2
Oceania	101	12.6	45	4.3	180	7.2	176	13.9	18	2.3	53	5.2	261	10.4	117	9.3
North America	277	50.8	257	86.1	736	51.6	1021	32.7	1157	18.9	2172	51.1	956	43.7	757	25.7
Latin America	5214	27.5	6649	20.2	6786	20.4	4798	18.6	1152	6.1	2958	9.0	2476	7.4	1622	6.3
EEC (9)	87	4.7	158	4.7	433	13.2	418	13.6	29	1.6	74	3.2	89	2.7	33	1.1
Northern Europe	63	10.1	145	9.4	656	13.9	524	8.5	20	3.3	82	5.3	241	5.1	227	3.7
Southern Europe	31	2.4	35	2.0	24	1.0	74	2.6	20	1.5	19	1.1	22	0.9	33	1.1
Central Europe	184	3.0	293	3.9	329	3.4	74	0.8	205	3.3	109	1.4	118	1.2	145	1.5
Eastern Europe	-	-	-	-	-	-	-	-	173	2.3	232	3.4	400	5.4	430	5.7
USSR	-	-	-	-	-	-	-	-	-	-	-	-	-	-	-	-
World	12009	19.2	15134	16.4	16841	16.5	12688	12.8	4365	7.0	10200	11.0	7084	6.9	5008	5.1

In 1970, 27.5 per cent of the EEC steel exports was imported by the region. But, between 1970 and 1982, a downward trend was recorded in this proportion and, by 1982, the share had dropped to 18.6 per cent.

Africa sent a considerable proportion of its steel exports to North America, but this proportion underwent significant fluctuation, particularly between 1970 and 1978. The share of African steel exports imported by the region, which accounted for 26.1 per cent in 1970, dropped to 5.9 per cent in 1974 and remained at around 10 per cent up to 1977. Amounting to 49.9 per cent in 1978, the share was more or less stable until 1982 (at around 50 per cent).

Between 1970 and 1974, the Far East recorded a significant increase in its share of steel exports imported by the region: it grew from 8.8 per cent to 57.0 per cent. During the next years there was a downward trend, with the share dropping to 13.6 per cent by 1982.

The share of the Oceanian steel products imported by the region saw significant changes between 1970 and 1982. Between 1970 and 1981, it decreased from 12.6 to 2.6 per cent and by 1982 it had grown to 13.9 per cent. The shares of the southern and the central countries of western Europe were more or less stable, accounting for around 10 per cent and 2.5 per cent, respectively. The share of the east European countries' steel products imported by the region fluctuated at around 3.5 per cent between 1970 and 1978 and, by 1982, had dropped to 0.8 per cent.

Latin America

Between 1970 and 1982, the share of exports to Latin America in the total steel exports of the North American countries was larger than that of any of the other main exporting regions of the world. During the period from 1970 to 1974, it increased from 18.9 to 51.1 per cent. After 1974 there was a downward trend, and the share dropped to 25.7 per cent by 1982 (see table 3.7).

Other exporting regions with an interest in the steel market of the Latin American countries were the following:

- Africa: the share of African steel products imported by the region dropped to 1 per cent in 1971, compared with 10 per cent in 1970; it subsequently grew to 16.3 per cent in 1975 and dropped again to 5.9 per cent by 1982;

- Oceania: the share underwent considerable changes between 1970 and 1982. Accounting for 2.3 per cent in 1970, it increased to 23.4 per cent by 1975 and then dropped to 1.4 per cent in 1980, following an increase in 1982 (9.3 per cent);

- Japan: the share of Japanese steel products imported by the region grew from 9.0 per cent in 1970 to 13.6 per cent in 1974. This was followed by a downward trend and a decrease to 5.2 per cent in 1982;

- EEC: in 1975 the share was at its maximum level, amounting to 10.3 per cent. During the rest of the period the share remained more or less stable, between 6 and 8 per cent;

- Far East: between 1970 and 1973, the share of steel exports imported by the region increased from 0.2 to 8.4 per cent, followed by a downward trend, when it fell to 0.2 per cent in 1978; it grew again in 1979 (3.6 per cent) and then dropped to 1.5 per cent in 1982; and

- The USSR: the share increased from 2.3 per cent in 1970 to 5.7 per cent in 1982. Attention is, however, drawn to the fact that, for the purpose of the present study, the volume of USSR steel exports for the period from 1978 to 1982 is estimated.

The shares of other exporting regions were rather small and recorded no significant change.

Middle East

The changes in the relative importance of the Middle East as an importer of steel products for various exporting regions are presented in table 3.8. The main changes which occurred in the relative importance of this importing region between 1970 and 1982 were the following:

- Japan considerably increased its export share to the region, from 3.7 to 20.3 per cent;

- Even stronger growth was recorded by the southern countries of western Europe. Their export share to the region increased from 5.7 per cent to 31.8 per cent;

- The Latin American countries were exporting a rather large proportion of their steel exports to the region by 1982, with a share reaching 16.9 per cent, compared with 1.5 per cent in 1970;

- The North American countries considerably increased the proportion of their steel products imported by the region. The share grew from 1.2 per cent in 1970 to 21.4 per cent in 1982; and

- In contrast with the above exporting regions, the shares of steel exports of the Far East and the USSR dropped from 27.9 per cent to 15.3 per cent and from 5.9 to 1.0 per cent, respectively.

The proportion of steel exports from EEC to the region grew from 5.3 per cent in 1970 to 12.5 per cent in 1978; this was followed by a decrease, with the share dropping to 10.1 per cent by 1982.

Oceania, the northern and the central countries of western Europe and the east European countries increased the shares of their steel products imported by the region from 1.4 to 7.5 per cent, from 0.7 to 4.2 per cent, from 2.4 to 4.2 per cent and from 9.6 to 11.5 per cent, respectively.

Between 1970 and 1977, the proportion of African steel exports to the region grew from 11.4 to 40.8 per cent, but, since 1979, Africa has exported practically no steel products to this destination.

Table 3.8. Imports of steel products: Middle East

| Years | Middle East | | | | | | | |
| | 1970 | | 1974 | | 1978 | | 1982 | |
Exporters	1 000 t	%	1 000 t	%	1 000 t	%	1 000 t	%
Africa	26	11.4	132	38.4	36	2.8	0	0.0
Far East	277	27.9	149	9.9	501	17.4	1 036	15.3
Japan	608	3.5	3 223	10.0	3 733	12.1	5 810	20.3
Oceania	11	1.4	20	1.9	189	7.5	47	3.7
North America	72	1.2	280	6.6	199	9.1	629	21.4
Latin America	8	1.5	5	1.8	50	3.5	529	16.9
EEC (9)	1 010	5.3	2 768	8.4	4 175	12.5	2 609	10.1
Northern Europe	12	0.7	36	1.6	119	3.6	130	4.2
Southern Europe	36	5.7	175	11.3	673	14.2	1 947	31.8
Central Europe	31	2.4	72	4.1	31	1.3	120	4.2
Eastern Europe	589	9.6	549	7.3	793	8.1	1 128	11.5
USSR	435	5.9	238	3.5	68	0.9	75	1.0
World	3 116	5.0	7 646	8.3	10 568	10.4	14 061	14.2

3.3 Examination of structural changes in total imports of steel products imported by separate countries and/or regions broken down by type of product irrespective of the place of origin

The European Economic Community

The changes which took place during the period in the product pattern of EEC imports are presented in part A of table 3.9. The main feature of that product pattern is the increase recorded in the groups of long products and flat products. The shares of these groups, accounting, in 1970, for 16.5 and 36.6 per cent, respectively, of total EEC steel imports, grew to 31.1 and 40.7 per cent.

The other trend brought out by the table is a considerable falling-off in the proportion of the group of ingots and semis. The share of this group decreased from 41.3 per cent in 1970 to 17.7 per cent in 1982.

The share of the group of tubes and fittings had increased from 4.7 to 11.1 per cent between 1970 and 1974, by 1978 it had dropped to 8.1 per cent and from then until 1982 it had fluctuated around 8.0 per cent.

The group of wire accounted for only a small proportion of the total EEC imports of steel products, its share increasing slightly, from 0.9 to 2.3 per cent, during the period.

North European countries

Part B of table 3.9 shows the product composition of steel imports of the North European countries and the changes in the product pattern during the period.

The most important development was noted in the group of flat products, whose share in the total steel imports of northern Europe was 54.3 per cent in 1970; by 1978 it had dropped to 51.7 per cent and in 1982 it recorded a slight increase (53.3 per cent).

Another important group of products was long products. However, during the period under review, its share showed a downward trend, declining from 28.2 per cent in 1970 to 20.4 per cent in 1982.

An increase occurred for the group of tubes and fittings, with the share growing from 10.8 to 17.1 per cent between 1970 and 1982.

The group of wire constituted the smallest category of imported products, its share dropping from 1.6 to 1.0 per cent during the period.

The share of the group ingots and semis increased from 5.0 to 8.1 per cent between 1970 and 1982.

South European countries

One of the most striking developments in the structure of steel product imports of the South European countries is the growth of the group of ingots and semis (part C of table 3.9). The share of this group, which accounted for 25.8 per cent in 1970, grew to 46.5 per cent by 1982.

A rather considerable falling off may be noted in the proportion of the group of flat products. In 1970 the share of this group in total steel imports of the South European countries was 51.1 per cent, in other words more than half of the region's steel imports. Between 1970 and 1982 the share declined to 33.4 per cent.

The group of long products made up considerable proportion of the total steel imports of the region. Its share decreased slightly, from 18.3 to 15.0 per cent, between 1970 and 1982.

The smallest proportion was accounted for by the group of wire. During the period, the share of this group did not change significantly, fluctuating between 1.2 and 1.7 per cent.

The table points to the relative stability of the proportion of the group of tubes and fittings in steel imports of southern Europe. The share in question fluctuated between 3.2 and 3.8 per cent during the period.

Central European countries

It may be seen in part D of the table, that, in 1982, the group of flat products represented 41.6 per cent of the total steel imports of the central European countries, compared with 38.3 per cent in 1970.

The other main part of the region's steel imports was made up of the group of long products. With some fluctuation, the share of this group increased from 32.1 per cent in 1970 to 36.4 per cent in 1982.

Table 3.9. Structural changes of imports of steel products a/
(Percentage of region imports)

Products / Years	A. European Economic Community							
	1970		1974		1978		1982	
	1 000 t	%	1 000 t	%	1 000 t	%	1 000 t	%
Ingots and semis	3 502	41.3	1 823	28.4	2 603	28.8	1 601	17.7
Long products	1 398	16.5	1 145	17.8	2 304	25.5	2 812	31.1
Flat products	3 100	36.6	2 621	40.8	3 177	35.2	3 682	40.7
Wire	79	0.9	116	1.8	213	2.4	207	2.3
Tubes & fittings	402	4.7	715	11.1	735	8.1	738	8.2
Total	8 481	100.0	6 421	100.0	9 031	100.0	9 040	100.0

Products / Years	B. Northern countries of western Europe							
	1970		1974		1978		1982	
	1 000 t	%	1 000 t	%	1 000 t	%	1 000 t	%
Ingots and semis	164	5.0	226	5.5	208	7.7	307	8.1
Long products	922	28.2	1 139	27.8	629	23.3	773	20.4
Flat products	1 772	54.3	2 164	52.8	1 397	51.7	2 014	53.3
Wire	53	1.6	55	1.4	41	1.5	38	1.0
Tubes & fittings	353	10.8	517	12.6	425	15.8	648	17.1
Total	3 264	100.0	4 102	100.0	2 700	100.0	3 781	100.0

Products / Years	C. Southern countries of western Europe							
	1970		1974		1978		1982	
	1 000 t	%	1 000 t	%	1 000 t	%	1 000 t	%
Ingots and semis	1 293	25.8	1 727	26.7	2 316	43.0	2 669	46.5
Long products	917	18.3	1 535	23.8	954	17.7	860	15.0
Flat products	2 563	51.1	2 885	44.7	1 832	34.0	1 913	33.4
Wire	69	1.4	102	1.6	90	1.7	71	1.2
Tubes & fittings	170	3.4	208	3.2	191	3.5	221	3.8
Total	5 013	100.0	6 457	100.0	5 382	100.0	5 734	100.0

a/ For the calculation of the shares of particular groups of products in steel imports, in tables E, G and I, no account could be taken of imports of countries which had not provided data broken down by product.

The shares of wire and tubes and fittings varied relatively little over the period. The share of wire increased from 1.6 to 2.6 per cent and the share of tubes and fittings, which amounted to 11.4 per cent in 1970, increased to 12.4 per cent by 1974 and dropped again, to 10.5 per cent, in 1982 (part D, table 3.9).

A considerable part of the central European countries' imports was in the form of ingots and semis. Their share accounted for 16.7 per cent in 1970, dropped to 12.4 per cent in 1974, grew to 16.7 per cent in 1978 and then declined again, amounting to 8.9 per cent in 1982.

East European countries

The structural changes between 1970 and 1982 in the product composition of steel imports of the east European countries are shown in part E of table 3.9. Owing to the lack of information from the Soviet Union about its exports to this region by category of steel products between 1978 and 1982, the relevant statistical data were estimated by the secretariat.

While the major part of the region's steel imports was made up of the group of flat products, the share of this group in total steel imports declined with some fluctuations from 51.5 per cent in 1970 to 43.7 per cent in 1982.

Another important group of products in the region's steel imports was long products, whose share, although fluctuating, increased from 30.6 to 36.6 per cent between 1970 and 1982.

The share of the group of ingots and semis varied relatively little during the period, increasing from 9.7 to 11.5 per cent. Between 1970 and 1978 the share of tubes and fittings grew from 6.8 to 10.7 per cent, followed by a decrease in 1982 (7.1 per cent).

Union of Soviet Socialist Republics

Changes which occurred between 1970 and 1982 in the product pattern of USSR steel imports are shown in part F of table 3.9. The composition of USSR imports from the German Democratic Republic and Romania for 1978 and 1982 was estimated by the secretariat.

As may be seen from the table, there were no significant structural changes. The data presented illustrate the rather high proportion of the group of tubes and fittings which, in 1970, accounted for 46.0 per cent. By 1974, this had dropped to 25.8 per cent and between then and 1982 the share remained relatively stable, fluctuating between 40.0 and 44.0 per cent.

Another important group of products was flat products, whose share accounted for more than one third of the total USSR steel imports between 1970 and 1982. The share nevertheless decreased slightly, from 37.9 per cent in 1970 to 36.3 per cent in 1980.

Table 3.9. Structural changes of imports of steel products (continued)
(Percentage of region imports)

D. Central countries of western Europe

Years / Products	1970 1 000 t	%	1974 1 000 t	%	1978 1 000 t	%	1982 1 000 t	%
Ingots and semis	392	16.7	299	12.4	451	16.7	217	8.9
Long products	752	32.1	779	32.3	822	30.4	888	36.4
Flat products	899	38.3	988	41.0	1 059	39.1	1 015	41.6
Wire	37	1.6	45	1.9	48	1.8	64	2.6
Tubes & fittings	266	11.4	299	12.4	326	12.1	256	10.5
Total	2 346	100.0	2 410	100.0	2 706	100.0	2 440	100.0

E. East European countries

Years / Products	1970 1 000 t	%	1974 1 000 t	%	1978 1 000 t	%	1982 1 000 t	%
Ingots and semis	676	9.7	775	9.1	939	11.0	823	11.5
Long products	2 124	30.6	3 288	38.5	2 938	34.3	2 618	36.6
Flat products	3 574	51.5	3 556	41.7	3 668	42.8	3 126	43.7
Wire	94	1.3	108	1.3	99	1.2	77	1.0
Tubes & fittings	473	6.8	803	9.4	914	10.7	511	7.1
Total	6 941	100.0	8 579	100.0	8 558	100.0	7 155	100.0

F. USSR

Years / Products	1970 1 000 t	%	1974 1 000 t	%	1978 1 000 t	%	1982 1 000 t	%
Ingots and semis	4	0.2	108	1.4	59	0.7	18	0.2
Long products	357	15.9	2 936	38.5	1 900	21.3	1 874	18.6
Flat products	853	37.9	2 574	33.8	3 189	35.7	3 656	36.3
Wire	1	0.1	33	0.4	173	1.9	108	1.1
Tubes & fittings	1 034	46.0	1 969	25.8	3 603	40.4	4 417	43.9
Total	2 250	100.0	7 619	100.0	8 924	100.0	10 072	100.0

The share of each of the other three groups of products showed the following changes:

- Ingots and semis: amounting to 0.2 per cent in 1970, the share increased to 1.4 per cent in 1974 and it dropped again to 0.2 per cent by 1982;

- Long products: in 1970, the share amounted to 15.9 per cent of total steel imports; this was followed by an increase in 1974 (38.5 per cent) and then a downward trend, with the share declining to 18.6 per cent by 1982;

- Wire: the share was insignificant, accounting for around 1.0 per cent for the main part of the period under review.

Far Eastern countries

During the period under review, the following changes occurred in the structure of the steel imports of the Far Eastern countries (part G of table 3.9):

- Accounting for 15.2 per cent of total steel imports in 1970, the share of ingots and semis increased to 24.3 per cent by 1974, i.e. almost one fourth of the region's steel imports. This was followed by a downward trend, and by 1982 the share had dropped to 15.2 per cent;

- The share of the group of long products underwent no significant changes between 1970 and 1982: it fluctuated between 20.0 and 25 per cent;

- The major part of the region's steel imports was made up of the group of flat products, which accounted for 50.5 per cent in 1970. After that the share remained relatively unchanged, amounting to 49.8 per cent in 1982;

- The group of wire made up the smallest proportion. Its share decreased slightly, from 2.7 to 0.7 per cent;

- The share of the group of tubes and fittings remained virtually unchanged, ranging from 9.5 to 11.0 per cent during the period.

Japan

The main feature of the product pattern of Japanese steel imports is the increase of the group of flat products, which in 1970 accounted for 17.3 per cent and by 1982 had grown to 49.3 per cent (part H of table 3.9).

The other trend brought out by the table is the very high proportion of the group of ingots and semis. In 1974 this share reached a peak of 79.2 per cent of total Japanese steel imports, compared with 49.8 per cent in 1970. The share then declined, dropping to 45.9 per cent by 1982.

A significant falling-off occurred in the share of the group of long products, with some fluctuation, it decreased from 25.8 per cent in 1970 to 3.0 per cent in 1982.

Table 3.9. Structural changes of imports of steel products (continued)
(Percentage of region imports)

	G. Far East							
Years	1970		1974		1978		1982	
Products	1 000 t	%	1 000 t	%	1 000 t	%	1 000 t	%
Ingots and semis	1 275	15.2	3 456	24.3	4 861	22.2	2 609	15.2
Long products	1 678	20.3	3 127	22.0	5 454	24.9	3 986	23.3
Flat products	4 179	50.5	5 951	41.8	9 339	42.6	8 524	49.8
Wire	226	2.7	165	1.2	178	0.8	119	0.7
Tubes & fittings	920	11.1	1 526	10.7	2 100	9.5	1 897	11.0
Total	8 278	100.0	14 225	100.0	22 026	100.0	17 391	100.0

	H. Japan							
Years	1970		1974		1978		1982	
Products	1 000 t	%	1 000 t	%	1 000 t	%	1 000 t	%
Ingots and semis	58	49.8	239	79.2	239	64.4	890	45.9
Long products	30	25.8	16	5.2	43	11.7	58	3.0
Flat products	20	17.3	21	7.0	66	17.9	957	49.3
Wire	1	1.0	6	2.2	8	2.2	1	0.0
Tubes & fittings	7	6.1	19	6.4	14	3.9	34	1.7
Total	117	100.0	301	100.0	371	100.0	1 941	100.0

	I. North America							
Years	1970		1974		1978		1982	
Products	1 000 t	%	1 000 t	%	1 000 t	%	1 000 t	%
Ingots and semis	1 166	9.7	1 696	11.2	2 692	16.0	1 727	13.6
Long products	3 528	29.4	5 208	34.4	3 615	21.5	2 719	21.4
Flat products	5 106	42.5	6 020	39.8	7 247	43.0	4 177	32.9
Wire	476	4.0	633	4.2	432	2.6	203	1.6
Tubes & fittings	1 735	14.4	1 577	10.4	2 855	17.0	3 861	30.4
Total	12 009	100.0	15 134	100.0	16 841	100.0	12 688	100.0

There was a downward trend in the proportion of the group of tubes and fittings, whose share decreased from 6.1 to 1.7 per cent between 1970 and 1982. During the same period, Japan practically ceased importing wire. Its share dropped from 1.0 per cent to zero.

North American countries

Steel product imports of the North American countries showed the following changes during the period (part I of table 3.9):

- The share of the group of ingots and semis grew from 9.7 to 16.0 per cent between 1970 and 1978, and dropped again to 13.6 per cent, by 1982;

- The share of the group of long products recorded its maximum of 34.4 per cent in 1974, compared with 29.4 per cent in 1970. This was followed by a decreasing trend, and by 1982 it had dropped to 21.4 per cent;

- The major part of the steel imports of the region was made up of the group of flat products, but the share of these products was not stable during the period. Accounting for 42.5 per cent in 1970, it had decreased to 39.8 per cent by 1974; in 1978 it recorded a slight increase (43.0 per cent) and dropped again to 32.9 per cent in 1982;

- As in all main regions, the group of wire constituted the smallest category of imported steel products, its share dropping from 4.0 to 1.6 per cent between 1970 and 1982;

- The progression of the group of tubes and fittings was particularly notable during the period, the share of the group increasing from 17.0 to 30.4 per cent. In 1970 the share accounted for 14.4 per cent.

Latin American countries

The major part of the steel imports of the Latin American countries was imported in the form of flat products (part J of table 3.9). But the significance of this group decreased: after amounting to 48.1 per cent in 1974, the share dropped to 36.3 per cent by 1982. The same trend was noted in the case of the group of ingots and semis. Occupying the second position after the group of flat products in 1974, when the share accounted for 23.9 per cent, the proportion decreased to 20.5 per cent by 1982.

There was also a downward trend in the share of the group of long products in the total steel imports of the region; it recorded a decline from 19.7 per cent in 1974 to 16.1 per cent in 1982.

The share of the group of tubes and fittings more than doubled during the period under review, from 10.5 to 26.0 per cent.

A rather low level may be noted in the proportion of the group of wire. The share of wire in total steel imports declined from 2.9 per cent in 1970 to 1.1 per cent in 1982.

Africa

The major share of the total steel imports of Africa, accounting for 36.6 per cent in 1970, was made up of the group of long products (part K of table 3.9). After 1970 there was an upward trend and by 1982 the share of these products had increased to 42.7 per cent.

Table 3.9. Structural changes of imports of steel products (continued)
(Percentage of region imports)

Products	J. Latin America							
Years	1970		1974		1978		1982	
	1 000 t	%	1 000 t	%	1 000 t	%	1 000 t	%
Ingots and semis	950	21.8	2 436	23.9	1 213	18.2	940	20.5
Long products	850	19.5	2 012	19.7	1 178	17.6	737	16.1
Flat products	1 983	45.4	4 907	48.1	2 481	37.1	1 663	36.3
Wire	125	2.9	178	1.7	82	1.2	50	1.1
Tubes & fittings	457	10.5	667	6.5	1 730	25.9	1 188	26.0
Total	4 365	100.0	10 200	100.0	7 084	100.0	5 008	100.0

Products	K. Africa							
Years	1970		1974		1978		1982	
	1 000 t	%	1 000 t	%	1 000 t	%	1 000 t	%
Ingots and semis	197	5.9	545	10.0	577	11.4	680	13.3
Long products	1 219	36.6	2 204	40.4	2 035	40.3	2 181	42.7
Flat products	1 050	31.5	1 733	31.7	1 311	25.9	1 284	25.2
Wire	125	3.8	175	3.2	165	3.3	136	2.7
Tubes & fittings	740	22.2	804	14.7	966	19.1	822	16.1
Total	3 332	100.0	5 461	100.0	5 053	100.0	5 103	100.0

Products	L. Oceania							
Years	1970		1974		1978		1982	
	1 000 t	%	1 000 t	%	1 000 t	%	1 000 t	%
Ingots and semis	24	2.7	311	19.6	55	6.7	96	7.9
Long products	219	24.8	247	15.6	108	13.1	186	15.3
Flat products	457	51.7	788	49.7	486	59.1	545	44.6
Wire	24	2.7	48	3.0	17	2.1	35	2.9
Tubes & fittings	160	18.1	191	12.1	157	19.1	358	29.3
Total	885	100.0	1 586	100.0	822	100.0	1 221	100.0

Another important group of products in the region's steel imports was flat products. However, during the period under review, the share of these products declined from 31.5 per cent in 1970 to 25.9 per cent in 1982.

Steel product imports of the African countries showed the following further changes during the period:

- The share of the group of ingots and semis grew from 5.9 to 13.3 per cent;

- The share of the group of wire decreased slightly, from 3.8 to 2.7 per cent;

- While the group of tubes and fittings occupied the third position after the groups of long products and flat products, their share declined from 22.2 to 16.1 per cent.

Oceania

There were considerable structural changes in Oceanian steel imports between 1970 and 1982. The main development was the increase in the share of the group of tubes and fittings from 18.1 to 29.3 per cent and the decrease in the share of the group of long products from 24.8 to 15.3 per cent (part L of table 3.9).

The major part of the region's steel imports consisted of flat products. Accounting for 51.7 per cent in 1970, the share dropped to 49.7 per cent by 1974, followed by an increase in 1978 (59.1 per cent) and a decline to 44.6 per cent in 1982.

There was an upward trend in the share of the group of ingots and semis between 1970 and 1974, when the share grew from 2.7 to 19.6 per cent. After that, between 1974 and 1982, the share was more or less stable, fluctuating between 6.0 and 8.0 per cent.

There were no significant changes in the share of the group of wire between 1970 and 1982. Forming the smallest proportion of the total steel imports of the region, it fluctuated around 3.0 per cent during the period under review.

Middle East

The main feature of the changes which occurred in the product pattern of the total steel imports of the Middle East was the increase in the share of the group of long products from 42.4 per cent in 1970 to 54.0 per cent in 1982 (part M of table 3.9). Another important group of steel products imported by the region was flat products. During the period that share, although fluctuating, dropped from 28.2 to 20.2 per cent.

During the same period, the shares of ingots and semis and wire declined from 14.4 to 7.9 per cent and from 2.1 to 1.8 per cent, respectively. For some years the share of wire was even smaller, for instance in 1978 it accounted for only 1.2 per cent.

The group of tubes and fittings occupied the third position after the groups of long products and flat products. Its share in total steel imports of the region increased from 12.9 to 23.4 per cent between 1970 and 1974. After that, however, there was a downward trend, and the share dropped to 16.1 per cent by 1982.

Table 3.9. Structural changes of imports of steel products (concluded)
(Percentage of region imports)

Years	M. Middle East							
Products	1970		1974		1978		1982	
	1 000 t	%	1 000 t	%	1 000 t	%	1 000 t	%
Ingots and semis	449	14.4	1 090	14.3	1 305	12.3	1 117	7.9
Long products	1 320	42.4	2 904	38.0	4 738	44.8	7 598	54.0
Flat products	879	28.2	1 753	22.9	2 661	25.2	2 836	20.2
Wire	65	2.1	112	1.5	125	1.2	247	1.8
Tubes & fittings	403	12.9	1 787	23.4	1 749	16.5	2 263	16.1
Total	3 116	100.0	7 646	100.0	10 568	100.0	14 061	100.0

CHAPTER IV.   ANALYSIS OF CHANGES IN FOREIGN STEEL TRADE BALANCES OF
DIFFERENT COUNTRIES AND/OR REGIONS

Tables 2.1, 3.1 and 4.1 bring out the point that net exports and imports
only represent about 50 per cent of total world exports and imports.   The
greater part of world exports are balanced by imports into exporting countries
or regions, i.e. as a rule, the major exporting countries and/or regions are
at the same time also the main importing zones.   This is particularly true of
EEC, the other west European countries and the east European countries.

Africa

It will be seen in table 4.1 that, in 1970, Africa's net imports amounted
to 3.11 million tonnes and represented 93.4 per cent of the region's total
imports of steel products.   In spite of the fact that, during the period under
review, the region's net imports increased in tonnage terms to
4.03 million tonnes (in 1981 - 5.57 million tonnes), there was a downward
trend in the share of net imports in total steel imports, which dropped, with
some fluctuation, to 79.0 per cent by 1982.   As for the North African
countries (Algeria, Libyan Arab Jamahiriya, Morocco, Tunisia), the situation
of their trade balances was almost constant.   These countries did not export
steel products outside the region and their net imports increased from
1.15 million tonnes in 1970 to 2.68 million in 1982.   It should be mentioned
that, in the second half of the 1970s, South Africa became a net exporter,
exporting around 1.0 million tonnes of steel products with negligible imports
in 1982.

Far Eastern countries
(excluding Japan)

The Far East occupied the second position amongst net importing regions
of the world.   Its net imports increased from 7.23 million tonnes in 1970 to
18.88 million in 1978.   After 1978 there was a decrease, with imports dropping
to 10.36 million tonnes by 1982.   That year, net imports accounted for
59.6 per cent of the total imports of the region, compared with 87.4 per cent
in 1970.   It should, however, be borne in mind that this improvement was
mainly due to the fact that, by the end of the 1970s, the Republic of Korea
had become a net exporting country.   From the figures given above, it is clear
that, in spite of some improvement, the Far East countries depended largely on
imports to satisfy their demand for steel.

Japan

As can be seen from table 4.1, between 1970 and 1982 Japan was the
world's largest net exporter of steel products.   Its net exports increased
from 17.35 million tonnes in 1970 to 35.79 million in 1976.   During the
remaining years of the period, they showed a decrease, dropping to
26.66 million tonnes by 1982, mainly due to an increasing trend in Japanese
steel imports from the Far Eastern countries.   In 1970 Japanese net exports of
steel products accounted for 99.3 per cent of the total steel exports of the
country;   by 1976 the ratio had grown to 99.7 per cent and by 1982 it had
again dropped to 93.2 per cent.

Table 4.1. Foreign steel trade balances of various
countries and/or regions
(Millions of tonnes)

Regions	1970	1974	1978	1982
Africa	3.11	5.12	3.76	4.03
Far East	7.23	12.44	18.88	10.36
Japan	-17.35	-31.80	-30.50	-26.66
Middle East	3.12	7.65	10.57	14.06
Oceania	0.08	0.55	-1.68	-0.04
North America	5.89	10.88	14.65	9.74
Latin America	3.82	9.90	5.66	1.88
EEC (9)	-10.51	-26.52	-24.30	-16.79
Northern Europe	1.43	1.82	-0.58	0.71
Southern Europe	4.39	4.91	0.65	-0.39
Central Europe	1.04	0.66	0.24	-0.45
Eastern Europe	0.81	1.03	-1.17	-2.66
USSR	-5.16	0.78	1.56	2.50

Middle East

Expressed in absolute quantities, during almost the whole of the period
under review, the third place among net importers of steel products was
occupied by the Middle East. The region's net imports grew from
3.12 million tonnes in 1970 to 14.06 million in 1982. Its net imports were
equal to its total steel imports, since the Middle East did not export steel
products during the period.

Oceania

Until 1975, Oceania was a net importer of steel products and, between
1970 and 1974, its net imports of steel products rose from 0.08 million tonnes
to 0.55 million. In 1974, the ratio of net imports to total steel imports of
the region was 34.8 per cent. But, in 1975, Oceania became a net exporter of
steel products, with its net exports accounting for 0.77 million tonnes; by
1978, this had increased to 1.69 million tonnes. This was borne out by the
increasing proportion of steel products exported. After 1979 the situation
changed somewhat, and Oceanian net exports declined steadily, amounting to
0.04 million tonnes in 1982.

North American countries

Between 1970 and 1982, the North American countries were amongst the world's largest net importers of steel products. Amounting to 5.89 million tonnes in 1970, the net imports of the region grew to 14.15 million tonnes in 1971 and then dropped to 7.77 million tonnes by 1975. After that and up to 1982, the region's net imports fluctuated, reaching their maximum in 1977 (15.31 million tonnes) and dropping again, to 9.74 million tonnes, in 1982. The largest ratio of net imports of steel products to total steel imports of the region was registered in 1977, when it was 90.6 per cent. During the major part of the period, the ratio fluctuated between 50 and 87 per cent.

Latin American countries

During the period, the Latin American countries were a net steel importing region, their imports, however, being characterized by significant fluctuations. Between 1970 and 1974, they showed significant growth, reaching a peak at 9.90 million tonnes in 1974, compared with 3.82 million in 1970. Starting in 1975 and up to 1981, the region's net imports underwent no significant changes, fluctuating between 8 and 5 million tonnes. In 1982, however, they declined to 1.88 million tonnes, a mere 49.2 per cent of the 1970 figure. Among the countries of this region, Brazil should be mentioned, in view of its contribution to improving the trade balance of the region, particularly in the last years of the period, when Brazil became a net steel exporter.

The European Economic Community

EEC occupied the second position (after Japan) among net exporting regions. During the whole period, the region was a net exporter of steel products. Between 1970 and 1974, its net exports of steel products had grown from 10.51 million tonnes to 26.52 million tonnes, dropping, however, to 13.22 million tonnes by 1976. By 1978 they had again increased, to 24.30 million tonnes and, in 1982, they declined once more, to 16.79 million tonnes.

North European countries

The North European countries were net importers of steel products between 1970 and 1976, with net imports fluctuating between 1.0 million tonnes (in 1971) and 1.82 million tonnes (in 1974). In 1977, the region became a net exporter, with negligible net exports (0.01 million tonnes) and it remained a net exporter for the next two years. Between 1980 and 1982 the situation was unstable: in 1980 the region was a net importer (0.24 million tonnes); in 1981 it became a net exporter (0.17 million tonnes); and in 1982 it became a net importer again (0.71 million tonnes). Of the individual countries included in this region (Finland, Norway and Sweden), only Norway remained a net importer of steel products during the whole period, with Finland and Sweden becoming net exporters as of 1977.

South European countries

The South European countries had for a long time been a typical steel importing region. In 1970 the region's net imports amounted to 4.39 million tonnes. Between 1970 and 1975, its net imports fluctuated between over 2 million tonnes and less that 5 million tonnes. After 1974 there was a strong decline and, in 1980, net imports of the region amounted to only 0.50 million tonnes. In 1981 the region moved from a net importer position to a net exporter position, with a marginal surplus of 0.70 million tonnes. In 1982 there was a slight decrease, when net exports dropped to 0.39 million tonnes. The greatest contribution to improving the trade balance of the region was made by Spain, whose relative importance as a steel exporter has tended to increase in recent years, with Spain, in fact, being net exporter of steel products among the countries of the region.

Central European countries

The Central European countries moved from a deficit to a marginal surplus during the period under review, this improvement being due particularly to Austria, which, between 1970 and 1982, was a net exporter of steel products. It should also be mentioned that, owing to the fact that exports of steel products from Switzerland had been growing almost continuously during the period, that country managed to reduce its net import balance. In 1970 the region's net imports amounted to 1.04 million tonnes and in 1973 it reached its peak (1.31 million tonnes). Between 1974 and 1978 net imports continued to fluctuate widely, between 0.02 million tonnes (in 1976) and 0.66 million tonnes (in 1974). In 1979 the region registered a surplus (0.28 million tonnes) and during the next three years the region had remained a net exporter.

East European countries

From 1970 to 1976, the East European countries were a net steel importing region, their net imports fluctuating at around 1 million tonnes. In 1977, the region became a net exporting region, when it registered a marginal surplus and, from then on to 1982, the region was a net exporter of steel products, with a surplus of 2.66 million tonnes in 1982.

Union of Soviet Socialist Republics

As can be seen from table 4.1, at the beginning of the period, the Soviet Union was a net exporter of steel products. In 1970, the country recorded its biggest surplus, amounting to 5.16 million tonnes. At that time, the USSR occupied third position, after Japan and EEC, amongst the net exporting regions of the world. But between 1970 and 1973 the situation changed considerably, when USSR net exports dropped to 1 million tonnes, and in 1974 the country became a net importer of steel products with a deficit of 0.78 million tonnes. By 1976 the country's net imports had increased to 2.03 million tonnes, followed by a decrease in 1977 (0.06 million tonnes). During the next four years the USSR net imports of steel products fluctuated between 1.56 and 2.00 million tonnes and, in 1982, they reached their peak of 2.50 million tonnes.

It is important to note that USSR steel exports and imports for the period from 1970 to 1982 have been estimated according to the methodology described in chapter I.

CHAPTER V.   ANALYSIS OF CHANGES IN SPECIALIZATION COEFFICIENTS
OF VARIOUS EXPORTING COUNTRIES AND/OR REGIONS IN
VARIOUS IMPORTING ZONES

Specialization coefficients offer a convenient means of analysing the structure of international trade. They reveal where an exporting country is concentrating its trading effort most, independently of the quantities exported. On the other hand, they also reveal which exporting countries are making proportionally the greatest market-penetration effort in a given importing region. Changes in the value of the coefficients over time can also be used to assess changes in strategy of various exporters.

The following ratio gives the specialization coefficient of an exporting country in an importing region:

As the numerator: the share accounted for by the importing region of the exports of the country in question;

As the demoninator: the share accounted for by the same importing region of the exports of all the countries surveyed.

For the sake of simplicity, this ratio is multiplied by 100.

As a theoretical example, if a zone accounts for 30 per cent of the exports of a country, but only 10 per cent of the exports of all world exporters, the specialization coefficient of the exporting country in respect of the importing zone concerned is

$$K = 100 \times \frac{0.3}{0.1} = 300$$

The corresponding specialization coefficient of another exporting country in respect of which the importing zone accounts for only 5 per cent of its exports will be

$$K = 100 \times \frac{0.05}{0.1} = 50$$

The degree of specialization in a zone is then easy to determine. If its specialization coefficient is over 100, the exporting country is quite strongly specialized. A coefficient less than or equal to 100 indicates weak specialization. A coefficient scale of the following kind can be drawn up:

$K > 2;$ $1.7 < K \leq 2;$ $1.4 < K \leq 1.7;$ $1 < K \leq 1.4;$ $K \leq 1$

Specialization coefficients for the years 1970, 1974, 1978 and 1982 have been calculated for all steel products, for long products, for flat products and for tubes. The results of these calculations are presented in tables II.1 to II.16 in annex II.

A vertical analysis of the evolution of the coefficients shows their values and reveals how they change over time for individual exporting countries. A horizontal reading gives the coefficients of various exporting countries in respect of each importing region, indicating the relative importance of the region for each one.

Analysis by exporting country

European Economic Community (EEC) exports of all products in 1982 were concentrated mainly on west European countries, which is largely explained by their geographic and cultural proximity. Between 1970 and 1982, this specialization tended to increase, except in the case of flat products and tubes in respect of north European countries (see table 5.1).

EEC exports show a high specialization co-efficient in respect of "other North American countries", "other countries in Oceania" and "other Latin American countries". This is explained by the existence of overseas territories - especially French and British - in these regions.

Table 5.1. EEC specialization coefficients in respect of western Europe

	1970			1982		
	N.W.E.	S.W.E.	C.W.E.	N.W.E.	S.W.E.	C.W.E.
All products	254	132	267	224	224	293
Long products	204	134	231	266	237	324
Flat products	266	131	298	198	177	291
Tubes	219	151	234	148	179	244

EEC specialization coefficients are also high for African countries, reflecting the existence of old historic, linguistic and economic ties. These coefficients changed little during the period. They show greater specialization in long products than in flat products and tubes when the African continent as a whole is considered, but in reality specialization is very great in long products in respect of southern Africa and in flat products in respect of the north.

The relative importance of the North American market for the EEC was the same in 1982 as in 1970, following a decline in the coefficients in 1974 and 1978. EEC specialization coefficients increased sharply in respect of South America, and especially the four newly-industrialized countries (from 71 in 1970 to 158 in 1982).

The USSR is an export destination for which the EEC specialization coefficient has increased, particularly for flat products, albeit in an irregular manner, depending on the product and the year.

The west European countries show strong specialization coefficients in respect of EEC, representing the counterpart of the flows observed in EEC exports and reflecting the existence of very old and close trading links. In the case of destinations other than EEC, the strong concentration of exports from south European countries on Africa (especially North Africa) and the Middle East is noteworthy; here too, geographic proximity and the existence of very old-established trade flows play an obvious part. The same factors

explain the high specialization coefficients of the Central west European
countries in respect of the east European centrally-planned-economy countries.

The highest specialization indices among member countries of the Council
for Mutual Economic Assistance (CMEA) relate to western Europe (south and
central), EEC and, of course, the USSR. The exports of the USSR in 1974, a
year in which the statistics were officially available in tonnage terms,
revealed a high degree of specialization in respect of the Middle East,
South America (Cuba) and, of course, the other centrally-planned-economy
countries of eastern Europe. The estimates made for 1982 give substantially
the same picture, except that the Middle East market is of considerably less
importance.

North American exports show very high specialization coefficients in
respect of South America, although the coefficients are also significantly
high for the Far East, the Middle East, EEC and southern Europe. However,
certain markets lost some of their relative importance during the period in
question, e.g. Japan (140 in 1970; 87 in 1974; 674 in 1978 and 25 in 1982),
whereas the relative importance of others, such as the Middle East, has
increased.

Latin America's specialization coefficients were particularly high for
North America and EEC, throughout the period, which is characterized by two
features, namely, a decline in the relative importance of the African market,
and the specialization that emerged in Japan after 1974 (limited to flat
products).

For Oceania, the highest specialization coefficients are in respect of
the Far East and Latin America - the regions that are geographically the
closest.

In 1982 the exports of Far Eastern countries (except Japan) show high
specialization coefficients in respect of Japan and Oceania - which is
explained by their geographic proximity - but also in respect of certain
Middle Eastern countries and North America. The coefficient is also very high
(228) for the north of western Europe (503 for flat products), although only
in 1982.

Japanese exports are highly specialized in respect of two regions,
namely, the Far East and Oceania, and to a lesser degree in respect of
southern Africa, the Middle East, and North and South America. Over the
1970-1982 period, they reveal a continuing high degree of specialization in
respect of the Far East, Oceania and southern Africa, increased specialization
in respect of the Middle East, and a decline in respect of North America. The
specialization coefficients reveal essentially the same geographical structure
for each type of product (long and flat products and tubes).

Africa's exports show high specialization coefficients in respect of
North America, Latin American countries other than the four recently
industrialized ones, EEC and, to a lesser degree, the southern countries of
western Europe.

The simplified world charts presented in figures 5.1 to 5.5 provide a
clearer picture of the specialization regions of the main exporters in 1970
and 1982.

Key to figures 5.1 to 5.5

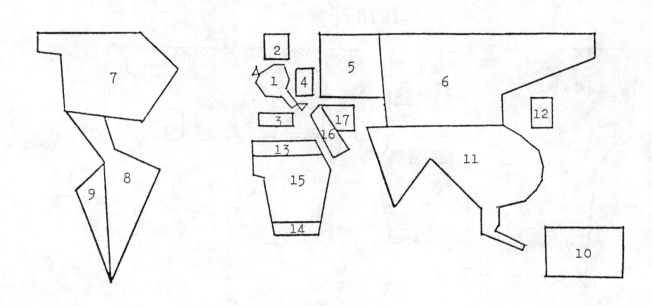

1.   The EEC (nine countries)

2.   Northern Europe

3.   Southern Europe

4.   Central Europe

5.   Centrally-planned-economy countries of eastern Europe without the USSR

6.   The USSR

7.   North America

8.   4 selected countries of Latin America (Argentina, Brazil, Mexico, Venezuela)

9.   Other Latin American countries

10. Oceania

11. Far East (without Japan)

12. Japan

13. North Africa (Algeria, Libyan Arab Jamahiriya, Morocco, Tunisia)

14. South Africa

15. Other Africa

16. 2 selected countries of Middle East (Egypt, Saudi Arabia)

17. Other Middle East

Figure 5.1. Specialization coefficients of EEC by region

1970

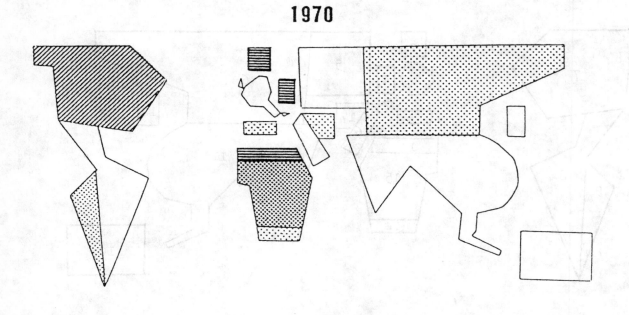

1982

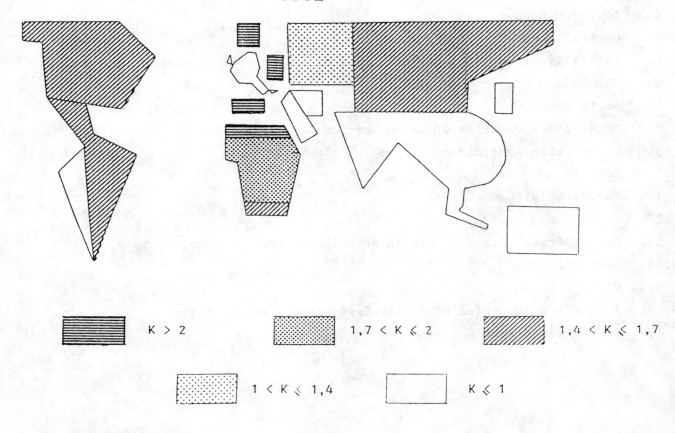

	K > 2		1,7 < K ⩽ 2		1,4 < K ⩽ 1,7
	1 < K ⩽ 1,4		K ⩽ 1		

Figure 5.2.   Specialization coefficients of Japan by region

1970

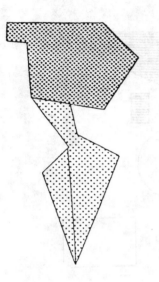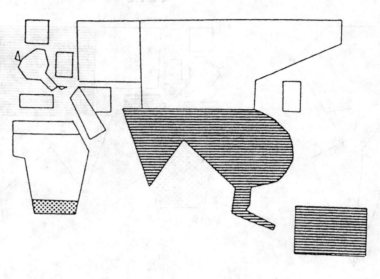

1982

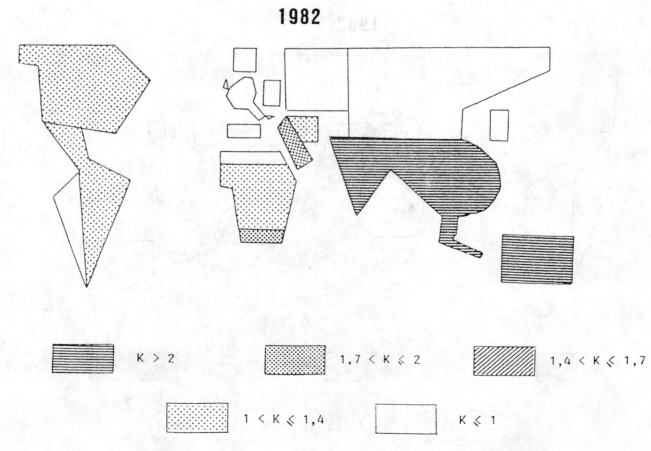

Figure 5.3.   Specialization coefficients of the Far East by region

1970

1982

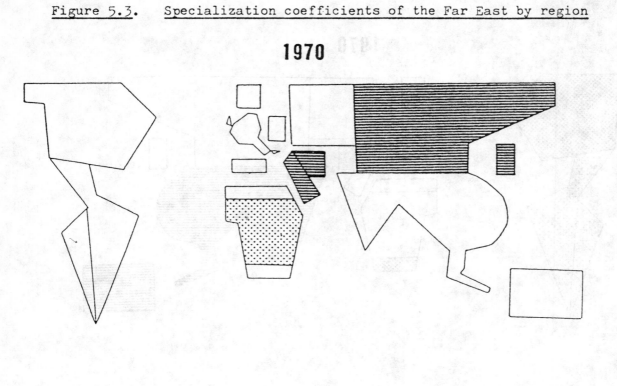
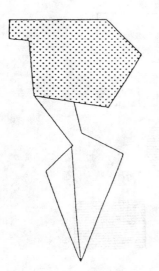
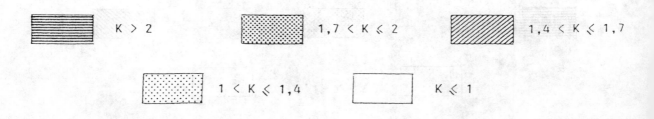

	K > 2		1,7 < K ⩽ 2	1,4 < K ⩽ 1,7

	1 < K ⩽ 1,4		K ⩽ 1

Figure 5.4. Specialization coefficients of Latin America by region

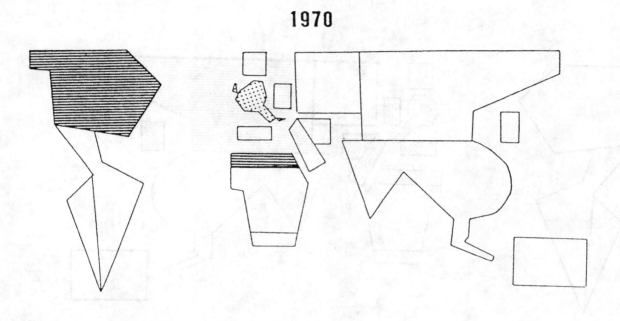

1970

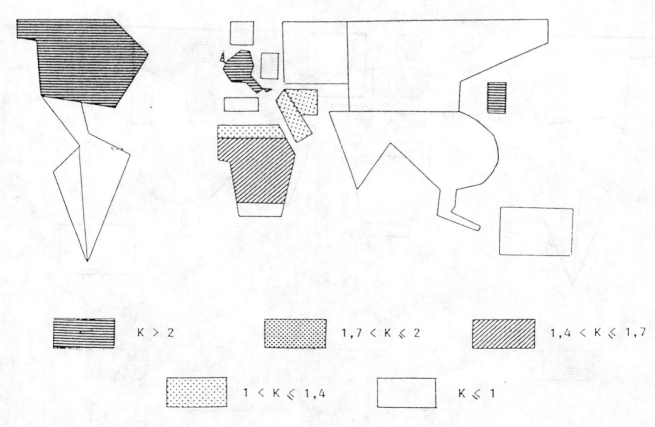

1982

Figure 5.5.   Specialization coefficients of East European countries

1970

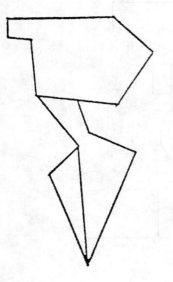
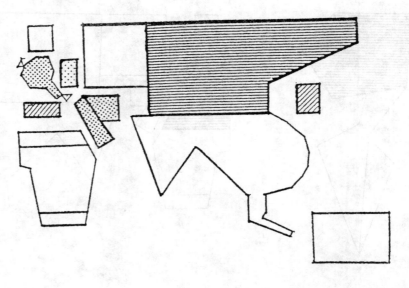

1982

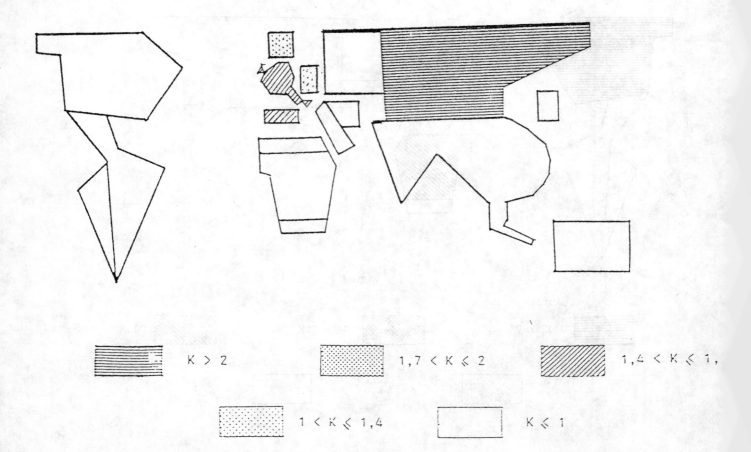

Analysis by importing region

Table 5.2, providing a comparison between regions and products, has been drawn up for the countries of Africa between 1970 and 1982.

Table 5.2. Countries with a specialization coefficient of over 100

	1970	1982
North Africa		
Long products	EEC Latin America	EEC Southern Europe Latin America
Flat products	EEC North America Latin America	EEC Southern Europe
Tubes	EEC Southern Europe Latin America	EEC Southern Europe Latin America
Southern Africa		
Long Products	EEC Japan	EEC Central Europe North America
Flat products	Japan	EEC Japan
Tubes	EEC Northern Europe Southern Europe North America Japan	Northern Europe North America Japan

North Africa retains its relative importance for EEC. In 1982, the interest of the south European countries in this market for long products increased whereas that of the Latin American countries in flat products declined somewhat.

The picture as regards southern Africa is more complicated, for this market is of relatively less importance to Japanese exporters of long products, and is one in which Central Europe and North America figure. Its importance to EEC is greater for flat products and less for tubes.

The Far East remains a market in which competition is mainly between countries bordering the Pacific, as may be seen in table 5.3.

Table 5.3. Countries with a specialization coefficient of over 100

	1970	1982
Long products	North America Oceania Japan	North America Oceania Japan
Flat products	North America Oceania Japan Africa	North America Latin America Oceania Other Far East Japan
Tubes	North America Japan	North America Japan Africa

Specializations in respect of the Middle East are numerous and more transient than in many other regions. Between 1970 (before the first oil shock) and 1982 (after the second oil shock), the market lost some of its relative importance for quite a number of exporters, particularly those of EEC with regard to flat and, later, long products. However, the region retains its relative importance for certain exporters, namely, those of southern Europe and the Far East in the case of long products, those of CMEA in the case of flat products, and those of North America, the Far East and (except in 1982) Japan in the case of tubes (see table 5.4).

Table 5.4. Countries with a specialization coefficient of over 100

	1970	1974	1978	1982
Long products	Southern Europe CMEA Far East	EEC Southern Europe CMEA Far East Japan Africa	EEC Southern Europe Far East	Southern Europe Latin America Far East Japan
Flat products	EEC Central Europe CMEA Oceania Far East Africa	Central Europe Japan Africa	CMEA Japan	Southern Europe CMEA North America
Tubes	North America Far East Japan	North America Far East Japan Africa	North America Far East Japan	Southern Europe North America Far East

The countries which are the most specialized in respect of North America often remain the same throughout the period. This is true in particular of EEC and Latin America. It is also worth mentioning that the relative importance of this market for Japanese exporters has declined (except as regards tubes), and that increased specialization by West European countries has emerged (see table 5.5).

Table 5.5. Countries with a specialization coefficient of over 100

	1970	1982
Long products	EEC Latin America Oceania Japan Africa	EEC Latin America Africa
Flat products	EEC Latin America Far East Japan	EEC Northern Europe Southern Europe Latin America Oceania Africa
Tubes	Latin America Oceania Japan	Latin America Far East Japan Africa

South America became a relatively important market for a number of exporting countries in 1982, although it is quite possible that this trend did not continue beyond that date. The Latin American market remains very important for North American exporters, which is not surprising, for European exporters, being a "traditional" market for flat and, more recently, long products, and for Japan, for flat products and tubes. It may be added that the relative importance of this market for African exporters has increased, as may be seen in table 5.6.

Table 5.6. Countries with a specialization coefficient of over 100

	1970	1982
Long products	EEC North America Oceania	EEC Central Europe North America Oceania Africa
Flat products	North America Japan	EEC Southern Europe North America Japan Africa
Tubes	Southern Europe USSR North America	EEC North America Japan

Western Europe remains the main market for European producers. Apart from European exporters, it may be noted that Africa is concentrating its sales on EEC and southern Europe, and that Far Eastern exporters had a high specialization coefficient in flat products in respect of northern Europe in 1982 (see table 5.7).

Table 5.7. Countries with a specialization coefficient of over 100

	1970	1982
Long products	EEC Northern, central and southern Europe CMEA North America Latin America	EEC Northern, central and southern Europe CMEA Africa
Flat products	Same as above, plus Africa	Same as in 1970, less southern Europe and plus the Far East
Tubes	Same as above	Same as in 1970

In the case of the centrally-planned-economy countries of eastern Europe, geographic proximity seems to be one of the main factors explaining specialization in this region; the specialization coefficients of west European countries, in particular, are higher for the countries situated geographically nearer to them than for the zone as a whole, as shown in table 5.8.

Table 5.8. Countries with a specialization coefficient of over 100

	1970	1982
Long products	Southern Europe Central Europe Far East	EEC Southern Europe */ Central Europe
Flat products	Southern Europe Central Europe	EEC Central Europe
Tubes	EEC Northern Europe	EEC Southern Europe */ Central and Northern Europe

*/ In respect of the region not including the USSR.

Notes

1/   During the period under review, Spain and Portugal were not members of EEC and Greece joined EEC only in 1981.

2/   Sources: United Nations Standard Country or Area Code for Statistical Use, Statistical Papers, Series M, No. 49/Rev.1, United Nations, New York, 1975 (ST/ESA/STAT/SER.M/49/Rev.1)

Customs Areas of the World, Statistical Papers, Series M, No. 30/Rev.1, United Nations, New York, 1970 ST/STAT/SER.M/30/Rev.1)

Annex I

DEFINITION OF INDIVIDUAL COMMODITIES IN ACCORDANCE WITH THE UNITED NATIONS
STANDARD INTERNATIONAL TRADE CLASSIFICATION (SITC)

1. Ingots and semis

672		Ingots and other primary forms of iron and steel (including blanks for tubes and pipes)
	672.1	Puddled bars and pilings, blocks, lumps and similar forms of iron and steel
	672.3	Ingots of iron and steel
	672.3(1)	- of other than high carbon or alloy steel
	672.3(2)	- of high carbon steel
	672.3(3)	- of alloy steel
	672.5	Blooms, billets, slabs, sheet bars and roughly forged pieces of iron or steel
	672.5(1)	- of other than high carbon or alloy steel
	672.5(2)	- of high carbon steel
	672.5(3)	- of alloy steel
	672.7	Iron or steel coils for re-rolling
	672.7(1)	- of other than high carbon or alloy steel
	672.7(2)	- of high carbon steel
	672.7(3)	- of alloy steel
	672.9	Blanks for tubes and pipes

2. Heavy sections

	673.4	Angles, shapes and sections (excluding rails), 80 mm or more, and sheet piling of iron or steel
	673.4(1)	- of other than high carbon or alloy steel
	673.4(2)	- of high carbon steel
	673.4(3)	- of alloy steel

3. Light sections

	673.2	Bars and rods (excluding wire rod) of iron or steel; hollow mining drill steel
	673.2(1)	- of other than high carbon or alloy steel
	673.2(2)	- of high carbon steel
	673.2(3)	- of alloy steel
	673.5	Angles, shapes and sections, less than 80 mm, of iron or steel
	673.5(1)	- of other than high carbon or alloy steel
	673.5(2)	- of high carbon steel
	673.5(3)	- of alloy steel

4. Plates

   674.1   Universals and heavy plates and sheets, more than 4.75 mm
                  in thickness, of iron or steel (other than tinned plates
                  and sheets)

      674.1(1)   - Heavy plates and sheets of iron or steel, other than of
                       high carbon or alloy steel (excluding tinned plates and
                       sheets)
      674.1(2)   - Heavy plates and sheets and universals of high carbon
                       steel
      674.1(3)   - Heavy plates and sheets and universals of alloy steel
      674.1(4)   - Universals of iron or steel, other than of high carbon
                       or alloy steel

   674.2   Medium plates and sheets, 3 mm to 4.75 mm in thickness, of
                  iron or steel (other than tinned plates and sheets)
      674.2(1)   - of other than high carbon or alloy steel
      674.2(2)   - of high carbon steel
      674.2(3)   - of alloy steel

5. Sheets, less than 3 mm

   674.3   Plates and sheets, less than 3 mm in thickness, of iron or
                  steel, uncoated
      674.3(1)   - of other than high carbon or alloy steel
      674.3(2)   - of high carbon steel
      674.3(3)   - of alloy steel

   674.8   Plates and sheets, less than 3 mm in thickness, of iron or
                  steel, coated (excluding tinned plates or sheets)
      674.8(1)   - of other than high carbon or alloy steel
      674.8(2)   - of high carbon steel
      674.8(3)   - of alloy steel

6. Hoops and strip

675   Hoop and strip of iron or steel

   675.0   Hoop and strip of iron or steel
      675.0(1)   - of other than high carbon or alloy steel
      675.0(2)   - of high carbon steel
      675.0(3)   - of alloy steel

7. Tinplate

   674.7   Tinned plates and sheets

8. Railway-track material

676   Rails and railway-track construction material of iron or
                  steel

   676.1   Rails of iron and steel
   676.2   Sleepers and other railway-track material of iron or steel

9.   <u>Wire rods</u>

    673.1   Wire rod of iron or steel
      673.1(1)   - of other than high carbon or alloy steel
      673.1(2)   - of high carbon steel
      673.1(3)   - of alloy steel

10.   <u>Wire</u>

    677   Iron and steel wire (excluding wire rod)

    677.0   Iron and steel wire (excluding wire rod)
      677.0(1)   - of other than high carbon or alloy steel
      677.0(2)   - of high carbon steel
      677.0(3)   - of alloy steel

11.   <u>Tubes and fittings</u>

    678.2   Tubes and pipes of iron (other than of cast iron) or steel,
               seamless (excluding clinched)
    678.3   Tubes and pipes of iron (other than of cast iron) or steel,
               welded, clinched etc.
    678.4   High pressure hydro-electric conduits of steel
    678.5   Tube and pipe fittings of iron or steel

12.   <u>Wheels, tyres and axles</u>

    731.7 <u>a</u>/   Wheels, tyres and axles (parts of railway locomotives and
               rolling-stock, n.e.s.)

<u>Notes</u>

    <u>a</u>/ A part only.

Annex II

STATISTICAL TABLES

The following abbreviations are used for the names of the regions.

Exporters

- E.E.C.   - Nine countries

- N.W.E.   - Northern Europe

- S.W.E.   - Southern Europe

- C.W.E.   - Central Europe

- C.M.E.A.   - Centrally-planned-economy countries of eastern Europe
                  without the USSR

- N. Amer   - North America

- L. Amer   - Latin America

- Oceania   - Australia, New Zealand

- F. East   - India, Republic of Korea

- Africa   - South Africa

Importers

- 3 selected countries of the Far East   - China, India, Republic of
                                                Korea

- 2 selected countries of the Middle East -   Egypt, Saudi Arabia

- 4 selected countries of Other America   - Argentina, Brazil, Mexico,
                                              Venezuela

- 7 selected countries of eastern Europe   - Albania, Bulgaria,
                                              Czechoslovakia, German
                                              Democratic Republic,
                                              Hungary, Poland, Romania

Table II.1. Coefficients of specialization of exporting regions for all products in 1970

Importers	E.E.C	N.W.E	S.W.E	C.W.E	C.M.E.A	U.R.S.S	N.Amer	L.Amer	Oceania	F.East	Japan	Africa	World
Africa	190.3	15.7	138.1	12.8	32.4	13.9	62.5	447.7	26.6	65.4	85.0	·	100.0
North Africa	210.1	14.7	122.6	14.0	42.7	31.7	54.3	1295.6	0.0	21.8	32.9	·	100.0
South Africa	112.5	39.9	13.6	18.8	0.0	0.0	132.3	2.9	17.8	0.0	181.8	·	100.0
Other Africa	196.8	10.4	179.6	10.5	33.8	5.6	50.3	0.1	46.4	110.5	94.9	16.3	100.0
Far East	47.9	24.3	28.8	5.8	26.0	19.3	102.3	0.0	440.9	48.0	225.1	0.0	100.0
3 selected	64.3	42.8	69.4	11.1	50.6	14.6	70.7	0.0	21.3	·	230.1	0.2	100.0
Japan	15.4	124.4	0.0	15.5	142.1	0.0	139.6	3.0	731.2	3419.0	0.0	27.6	100.0
Oth Far East	37.7	9.5	2.3	2.0	6.8	22.9	122.5	0.0	714.9	·	227.0	228.6	100.0
Middle East	106.6	13.5	115.0	47.5	192.6	117.7	23.7	30.9	27.5	560.0	69.8	0.0	100.0
2 selected	74.3	1.8	206.0	12.3	180.9	245.3	12.4	0.8	0.0	751.6	53.7	280.1	100.0
Oth Mid East	113.8	16.2	94.5	55.4	195.2	89.0	26.3	37.7	33.7	516.9	73.4	143.2	100.0
Oceania	78.0	37.3	8.5	14.5	0.0	0.0	64.0	1.7	·	53.4	240.0	0.0	100.0
Oth Oceania	254.7	0.8	0.7	0.1	0.0	0.0	17.6	0.0	·	0.4	74.3	135.7	100.0
N. America	142.8	24.7	52.5	12.2	15.6	0.0	·	264.2	65.5	45.7	175.8	0.0	100.0
N.Am unalloc	326.2	6.7	2.2	26.4	0.0	0.0	·	0.0	0.0	1.4	0.0	143.2	100.0
Oth America	86.8	22.5	47.3	22.1	47.8	33.4	270.6	·	32.3	2.8	128.2	237.6	100.0
4 selected	70.8	33.9	29.7	17.1	59.7	2.8	320.3	·	9.4	3.0	136.6	19.0	100.0
O Oth America	107.8	7.6	70.4	28.7	32.1	73.7	205.1	·	62.4	2.6	117.3	157.0	100.0
West Europe	111.4	272.0	146.3	240.5	156.4	23.7	196.9	77.1	64.9	11.7	42.5	200.0	100.0
EEC(9)	·	569.2	302.4	476.8	188.6	13.2	355.9	108.1	89.1	17.3	43.8	3.5	100.0
N.W.Europe	253.9	·	26.7	95.2	77.3	24.9	28.2	35.0	57.3	0.0	22.0	255.0	100.0
S.W.Europe	131.6	50.7	52.1	52.1	167.9	51.5	118.7	89.8	59.1	14.1	73.2	12.7	100.0
C.W.Europe	267.2	60.3	45.1	·	129.6	0.5	32.8	0.0	0.0	0.0	1.7	0.0	100.0
W.Eur unalloc	240.9	11.7	504.0	27.1	64.8	28.1	1.3	2.0	76.3	74.5	31.5	0.0	100.0
East Europe	62.3	32.2	166.6	123.1	79.2	508.0	9.7	0.0	23.7	95.3	18.1	0.0	100.0
7 selected	37.0	21.8	173.7	95.1	·	672.8	5.8	0.0	31.4	1.6	12.8	0.0	100.0
USSR	140.0	64.4	144.5	209.5	323.6	0.0	21.8	0.0	0.0	384.2	34.6	0.0	100.0
E.Eur unalloc	210.3	382.6	725.3	733.4	593.8	0.0	23.0	0.0	0.0	0.0	0.0	0.0	100.0
Unallocated	2.4	0.1	1.1	0.2	·	200.0	0.0	0.1	0.0	1083.3	0.0	0.0	100.0
World	100.0	100.0	100.0	100.0	100.0	100.0	100.0	100.0	100.0	100.0	100.0	100.0	100.0

Table II.2. Coefficients of specialization of exporting regions for long products in 1970

Importers	E.E.C	N.W.E	S.W.E	C.W.E	C.M.E.A	U.R.S.S	N.Amer	L.Amer	Oceania	F.East	Japan	Africa	World
Africa	175.3	14.1	63.9	14.1	53.2	23.3	95.6	576.8	39.0	44.5	33.8	.	100.0
North Africa	138.2	26.7	77.2	4.6	63.9	50.9	59.6	1484.7	0.0	27.7	4.2	.	100.0
South Africa	163.2	27.2	0.0	48.1	0.0	0.0	100.0	5.0	1.1	0.0	208.9	.	100.0
0ther Africa	205.6	2.2	65.8	14.9	55.0	6.8	121.9	0.0	75.5	65.4	23.2	.	100.0
Far East	52.1	37.7	84.9	30.8	16.5	20.4	177.2	0.0	494.4	37.8	408.1	8.5	100.0
3 selected	46.7	71.8	221.5	65.2	20.0	2.9	20.9	0.0	0.0	.	471.4	0.0	100.0
Japan	28.0	72.8	0.0	75.6	0.0	0.0	446.9	0.0	1466.9	2116.2	0.0	0.7	100.0
0th Far East	56.1	15.6	3.0	8.2	14.9	31.8	265.8	0.0	771.2	587.7	380.9	14.0	100.0
Middle East	83.6	17.1	131.8	45.0	202.8	95.0	16.3	38.6	0.0	841.8	8.1	4.7	100.0
2 selected	70.9	0.3	287.1	26.9	157.6	111.5	2.1	0.0	0.0	530.4	10.9	0.0	100.0
0th Mid East	86.5	20.9	96.7	49.1	213.0	91.2	19.5	47.3	0.0	.	7.4	5.7	100.0
Oceania	110.0	47.7	0.0	77.2	0.0	0.0	260.9	0.0	0.0	31.1	272.6	417.5	100.0
0th Oceania	238.4	0.0	0.0	0.2	0.0	0.0	8.9	0.0	.	0.0	53.8	.	100.0
N. America	153.5	37.7	5.2	16.0	29.3	0.0	.	117.2	175.9	49.2	199.1	333.2	100.0
N. Am unalloc	260.1	3.7	0.0	2.1	0.0	0.0	.	0.0	0.0	0.0	0.0	.	100.0
0th America	116.1	27.1	37.3	86.4	90.6	56.3	554.5	.	104.3	0.2	44.5	30.7	100.0
4 selected	81.6	51.9	24.3	140.1	148.5	86.3	869.0	.	0.0	0.0	44.9	0.6	100.0
10 0th America	134.5	13.8	44.2	57.8	59.7	4.9	386.8	105.3	160.0	0.4	44.3	46.8	100.0
West Europe	122.7	323.7	158.7	248.2	173.0	0.0	155.0	93.9	0.0	18.0	11.5	69.3	100.0
EEC(9)	.	880.3	421.0	549.3	268.2	20.9	322.3	52.3	0.0	37.2	7.6	198.3	100.0
N.W.Europe	203.6	67.9	36.0	75.9	101.1	0.8	34.5	269.4	0.0	0.1	0.7	2.7	100.0
S.W.Europe	133.7	24.1	.	181.8	189.2	0.0	155.3	0.0	0.0	22.3	36.2	1.1	100.0
C.W.Europe	231.0	3.7	5.7	.	72.7	0.0	3.7	0.0	0.0	0.0	0.0	0.0	100.0
W.Eur unalloc	208.9	5.4	279.6	8.5	69.6	0.0	1.4	0.0	0.0	6.9	32.7	0.0	100.0
East Europe	7.5	5.5	203.5	126.4	52.3	476.4	0.0	0.0	0.0	180.1	1.9	0.0	100.0
7 selected	8.7	4.6	146.0	147.6	.	556.6	0.0	0.0	0.0	0.6	0.0	0.0	100.0
USSR	0.6	227.4	545.8	0.0	363.2	0.0	0.0	0.0	0.0	1247.3	13.0	0.0	100.0
E.Eur unalloc	162.8	0.1	1083.2	873.4	383.4	0.0	0.0	0.0	0.0	0.0	0.0	0.0	100.0
Unallocated	3.7	.	1.4	0.2	.	268.1	0.1	0.0	0.0	0.0	0.0	0.0	100.0
World	100.0	100.0	100.0	100.0	100.0	100.0	100.0	100.0	100.0	100.0	100.0	100.0	100.0

Table II.3.   Coefficients of specialization of exporting regions for flat products in 1970

Importers	E.E.C	N.W.E	S.W.E	C.W.E	C.M.E.A	U.R.S.S	N.Amer	L.Amer	Oceania	F.East	Japan	Africa	World
Africa	154.6	8.7	82.6	11.7	6.9	0.6	37.1	38.1	102.6	0.6	135.7	.	100.0
North Africa	194.6	6.4	15.0	18.1	10.1	1.6	107.2	152.4	0.0	0.0	89.1	.	100.0
South Africa	91.1	36.9	0.0	4.2	0.0	0.0	10.6	0.0	66.6	0.0	194.1	.	100.0
Other Africa	153.7	2.6	131.2	11.0	7.3	0.3	14.5	0.0	154.2	0.9	140.5	.	100.0
Far East	56.6	29.4	1.1	1.0	38.2	14.7	137.2	0.0	214.2	0.5	177.6	114.6	100.0
3 selected	83.4	56.4	0.0	2.1	74.2	24.7	141.0	0.0	51.3		147.3	0.0	100.0
Japan	13.7	304.9	0.0	9.5	969.2	0.0	106.1	0.0	0.0	105.5	0.0	0.0	100.0
Oth Far East	35.9	5.8	1.9	0.2	1.9	7.0	134.4	0.0	343.9	.	202.9	205.6	100.0
Middle East	120.1	4.3	82.0	114.4	173.0	34.7	49.7	0.0	130.7	728.5	100.6	1238.5	100.0
2 selected	92.7	6.1	4.2	11.5	394.3	121.3	42.0	0.0	0.0	668.3	58.9	0.0	100.0
Oth Mid East	124.5	4.0	94.3	130.6	138.0	21.0	51.0	0.0	151.4	738.0	107.2	1434.5	100.0
Oceania	42.3	6.8	0.0	0.4	0.0	0.0	27.1	0.0	.	0.8	232.6	5.1	100.0
Oth Oceania	206.2	4.5	0.0	0.0	0.0	0.0	2.3	0.0		0.0	109.7	0.0	100.0
N. America	122.6	15.0	0.2	9.8	6.9	0.0	.	439.1	85.9	319.7	159.4	6.8	100.0
N.Am unalloc	344.0	31.7	0.0	0.0	0.0	0.0	201.1	0.0	0.0	152.1	0.0	0.0	100.0
Oth America	86.3	22.9	64.5	11.1	25.3	16.2	205.3	.	84.7	55.2	145.8	0.0	100.0
4 selected	99.2	35.7	87.6	4.2	33.8	0.0	194.4	.	25.7	62.8	138.8	0.0	100.0
10 Oth America	65.6	2.6	27.6	22.0	12.4	42.1	183.2		178.6	43.1	156.8	0.0	100.0
West Europe	129.3	274.6	162.6	199.9	180.8	15.1	332.0	51.0	82.6	38.7	53.2	132.1	100.0
EEC(9)	.	648.8	388.5	473.3	287.6	9.2	25.0	135.5	80.9	0.0	58.8	238.2	100.0
N.W.Europe	265.7	76.4	0.0	43.1	59.4	29.6	172.1	3.3	141.4	0.0	29.6	16.3	100.0
S.W.Europe	131.3	.	0.0	47.9	161.5	17.7	16.0	0.0	69.3	126.1	80.7	124.5	100.0
C.W.Europe	298.3	94.8	153.4	.	109.3	0.0	3.2	0.0	0.0	0.0	2.7	18.3	100.0
W.Eur unalloc	265.3	20.5	1094.2	157.5	8.7	0.0	25.0	3.5	755.5	0.0	19.7	0.0	100.0
East Europe	69.1	34.9	239.1	192.1	59.8	516.7	13.7	0.0	66.1	0.0	18.1	0.0	100.0
7 selected	49.1	41.5	290.7	100.1	.	640.1	72.5	0.0	81.8	0.0	14.0	0.0	100.0
USSR	152.6	7.1	22.8	577.2	310.3	0.0	42.8	0.0	0.0	0.0	35.1	0.0	100.0
E.Eur unalloc	227.7	549.5	1067.0	728.1	.	0.0	0.0	0.0	0.0	0.2	0.0	0.0	100.0
Unallocated	2.5	0.1	0.7	0.2	823.0	263.5	0.0	0.2	0.0	100.0	0.0	0.0	100.0
World	100.0	100.0	100.0	100.0	100.0	100.0	100.0	100.0	100.0	100.0	100.0	100.0	100.0

Table II.4. Coefficients of specialization of exporting regions for tubes in 1970

Importers	E.E.C	N.W.E	S.W.E	C.W.E	C.M.E.A	U.R.S.S	N.Amer	L.Amer	Oceania	F.East	Japan	Africa	World
Africa	176.7	35.2	181.7	31.4	1.7	18.3	192.8	118.0	0.0	227.3	54.6	.	100.0
North Africa	231.1	1.9	171.1	44.4	3.6	23.2	88.6	287.4	0.0	1.2	15.0	.	100.0
South Africa	103.9	210.2	103.9	79.2	0.0	0.0	130.7	7.4	0.0	0.0	136.8	.	100.0
Other Africa	142.9	42.3	197.8	16.6	0.3	16.5	278.6	0.7	0.0	422.9	76.1	0.0	100.0
Far East	31.5	23.1	3.4	2.1	19.2	19.7	158.7	0.0	43.8	0.9	219.8	0.0	100.0
3 selected	40.1	23.2	0.2	1.5	34.5	6.6	101.7	0.0	0.0	.	214.8	0.0	100.0
Japan	32.6	960.0	0.0	11.5	4.5	0.0	1487.0	0.0	0.0	109.8	0.0	0.0	100.0
Oth Far East	21.6	7.0	7.2	2.7	1.8	35.1	202.3	0.0	95.1	.	229.3	0.0	100.0
Middle East	97.0	10.6	75.9	9.7	23.6	63.1	103.2	4.6	0.0	810.6	126.9	0.0	100.0
2 selected	79.7	2.2	30.0	16.2	59.4	0.0	66.6	6.9	0.0	513.2	155.3	0.0	100.0
Oth Mid East	102.5	13.2	90.4	7.7	12.3	83.0	114.7	3.9	0.0	904.1	117.9	0.0	100.0
Oceania	75.6	108.4	2.2	0.8	0.0	0.0	116.4	13.1	.	335.2	172.7	0.0	100.0
Oth Oceania	163.7	0.0	2.5	0.0	0.0	0.0	155.8	0.0	.	0.0	94.5	0.0	100.0
N. America	74.6	25.0	32.7	38.3	2.5	0.0	.	229.2	407.6	55.6	188.2	38.4	100.0
N.Am unalloc	271.3	0.0	12.6	21.7	0.0	0.0	.	0.0	0.0	0.0	0.0	0.0	100.0
Oth America	80.5	74.1	137.9	40.9	4.1	131.8	703.8	.	0.4	4.5	86.6	0.0	100.0
4 selected	70.0	190.1	79.6	14.1	0.2	0.0	964.9	.	0.0	0.0	79.3	0.0	100.0
O Oth America	86.0	13.5	168.4	54.8	6.2	200.7	567.2	214.1	0.5	6.8	90.4	0.0	100.0
West Europe	139.4	363.1	261.9	480.7	136.0	6.0	60.4	642.0	0.0	71.6	21.3	564.8	100.0
EEC(9)	.	949.0	639.5	851.9	216.5	0.0	126.3	642.0	0.0	187.9	47.8	1082.1	100.0
N.W.Europe	219.4	118.5	53.8	628.4	52.9	7.3	9.0	0.9	0.0	0.0	5.6	610.2	100.0
S.W.Europe	151.4	.	.	91.5	237.0	9.6	105.9	0.0	0.0	1.1	19.9	184.1	100.0
C.W.Europe	233.5	136.0	104.4	.	68.3	11.4	2.9	0.0	0.0	0.0	2.1	.	100.0
W.Eur unalloc	183.0	32.1	814.7	21.1	0.0	0.0	7.0	5.3	0.0	705.2	37.3	0.0	100.0
East Europe	133.7	108.4	94.8	35.5	173.8	411.5	2.6	0.3	0.0	8.9	22.3	0.0	100.0
7 selected	91.0	8.6	302.0	106.9	.	1311.1	8.1	0.9	0.0	28.5	10.7	0.0	100.0
USSR	153.2	154.0	0.0	2.8	253.3	0.0	0.0	0.0	0.0	0.0	27.6	0.0	100.0
E.Eur unalloc	168.1	457.7	492.2	1064.3	.	0.0	80.3	0.0	0.0	0.0	0.0	0.0	100.0
Unallocated	1.7	0.2	1.4	0.4	833.8	7.5	0.2	0.3	0.0	0.1	0.0	0.0	100.0
World	100.0	100.0	100.0	100.0	100.0	100.0	100.0	100.0	100.0	100.0	100.0	100.0	100.0

Table II.5. Coefficients of specialization of exporting regions for all products in 1974

Importers	E.E.C	N.W.E	S.W.E	C.W.E	C.M.E.A	U.R.S.S	N.Amer	L.Amer	Oceania	F.East	Japan	Africa	World
Africa	178.1	35.7	116.0	14.9	16.2	13.2	116.6	80.5	17.7	32.2	71.4	.	100.0
North Africa	211.9	28.0	218.7	24.2	21.5	28.9	59.6	62.3	3.4	12.5	36.3	.	100.0
South Africa	146.0	50.7	54.5	9.6	4.5	0.0	202.3	19.8	57.5	87.5	97.4	.	100.0
Other Africa	159.8	35.9	41.3	7.9	16.9	3.9	130.9	131.1	11.6	23.8	94.2	16.2	100.0
Far East	31.3	11.8	21.1	3.9	21.7	19.3	92.3	1.0	397.3	10.5	218.8		100.0
3 selected	37.0	14.5	37.9	4.1	38.6	14.2	10.6	0.0	45.8	.	232.0	0.0	100.0
Japan	10.4	60.9	0.5	10.5	4.7	0.0	86.9	13.8	7344.5	497.3	0.0	4.1	100.0
Oth Far East	26.9	7.1	6.1	3.5	6.3	25.0	169.9	1.4	437.9		215.6	32.1	100.0
Middle East	101.6	19.2	136.6	49.5	87.9	42.1	79.6	22.0	23.2	119.3	121.4	464.6	100.0
2 selected	105.6	7.0	181.6	21.1	75.2	132.5	87.3	0.0	0.0	75.5	108.2	0.0	100.0
Oth Mid East	100.8	21.5	128.0	54.9	90.3	24.9	78.1	26.2	27.7	127.7	123.9	553.5	100.0
Oceania	32.2	30.7	1.8	12.4	1.0	0.0	91.7	2.6	.	270.7	226.6	25.7	100.0
Oth Oceania	131.2	0.2	20.7	0.3	20.2	0.0	60.4	0.0		449.3	118.5	2.5	100.0
N. America	123.3	42.4	57.1	12.1	23.7	0.0	.	525.6	26.5	347.9	127.0	36.0	100.0
N.Am unalloc	274.5	6.3	2.8	7.4	0.0	0.0	.	4.8	163.9	1.0	0.0	0.0	100.0
Oth America	81.4	29.4	48.3	9.6	13.1	30.7	463.0	.	46.7	71.2	122.8	87.5	100.0
4 selected	80.0	35.4	55.3	8.4	9.6	0.0	501.5	.	58.1	86.2	124.4	116.1	100.0
O Oth America	85.6	11.3	27.1	13.4	23.5	123.4	347.0	32.6	12.6	25.8	117.9	1.7	100.0
West Europe	124.7	355.1	146.1	335.5	182.4	30.4	85.4	75.4	114.8	18.4	41.8	203.1	100.0
EEC(9)	.	989.0	404.1	827.4	269.8	2.6	168.4	4.7	350.0	36.3	47.3	554.1	100.0
N.W.Europe	221.0	69.4	19.1	143.7	77.7	17.3	20.4	21.0	0.2	0.4	27.7	4.9	100.0
S.W.Europe	146.9	66.0	.	103.8	164.1	78.1	76.8	0.0	0.2	19.5	59.4	61.7	100.0
C.W.Europe	235.1	22.1	52.3	.	132.3	0.9	5.0	0.0	0.1	0.2	7.5	2.0	100.0
W.Eur unalloc	106.3	48.2	302.0	21.6	674.3	5.0	5.6	0.9	3.6	0.5	1.1	0.0	100.0
East Europe	113.8	75.2	187.6	114.1	129.5	433.9	8.8	1.4	11.1	0.9	28.0	0.0	100.0
7 selected	79.7	18.3	196.8	121.9	275.3	816.2	10.5	0.4	21.0	1.3	8.7	0.0	100.0
USSR	152.8	2.9	178.5	106.2		0.0	6.9	2.5	0.0	0.3	49.8	0.0	100.0
E.Eur unalloc	0.8	0.2	4.0	4.4		1344.5	0.6	0.0	0.0	0.0	0.0	0.0	100.0
Unallocated	4.9		0.9	0.2	852.8	270.1	0.0	0.5	0.7	514.1	0.6	0.0	100.0
World	100.0	100.0	100.0	100.0	100.0	100.0	100.0	100.0	100.0	100.0	100.0	100.0	100.0

Table II.6. Coefficients of specialization of exporting regions for long products in 1974

Importers	E.E.C	N.W.E	S.W.E	C.W.E	C.M.E.A	U.R.S.S	N.Amer	L.Amer	Oceania	F.East	Japan	Africa	World
Africa	184.3	27.3	62.1	41.7	23.7	17.3	48.3	55.0	38.1	2.1	33.0	·	100.0
North Africa	198.9	18.0	93.9	63.5	35.8	31.6	1.8	0.0	65.7	0.0	1.6	·	100.0
South Africa	140.6	22.2	33.2	32.9	0.0	0.0	192.9	0.0	13.2	1.0	112.7		100.0
Other Africa	177.7	45.5	19.6	8.1	12.9	0.0	67.3	172.7	1.4	6.3	53.4		100.0
Far East	26.4	17.4	33.2	17.6	19.3	5.8	156.7	1.5	736.7	24.2	302.2	24.7	100.0
3 selected	17.5	26.3	79.0	21.4	35.2	0.1	22.0	0.0	21.8		326.5	0.0	100.0
Japan	69.0	280.1	4.0	188.0	27.0	0.0	920.9	11.1	0.0		0.0	0.0	100.0
Oth Far East	31.5	10.0	6.0	14.0	9.8	0.0	231.4	2.4	1171.7	4866.4	290.1		100.0
Middle East	103.1	9.2	157.8	86.4	116.6	39.1	33.8	0.0	46.5	192.5	116.4	39.7	100.0
2 selected	121.2	9.5	216.6	21.5	117.3	34.2	23.1	0.0	0.0	130.4	92.6	426.5	100.0
Oth Mid East	98.7	9.2	143.7	102.2	116.5	40.3	36.4	0.0	57.7	207.5	122.1	0.0	100.0
Oceania	48.6	67.3	0.3	114.2	1.5	0.0	190.5	0.0		3.6	265.4	529.6	100.0
Oth Oceania	111.1	0.1	0.1	0.0	0.0	0.0	21.0	0.0		5.5	188.4	81.7	100.0
N. America	121.8	58.7	28.1	14.5	34.6	0.0		490.4	20.5	403.4	127.8	4.6	100.0
N.Am unalloc	221.5	1.9	0.8	4.0	0.0	0.0	743.1	0.0	0.0	0.0	0.0	68.1	100.0
Oth America	83.7	28.2	102.2	54.4	19.6	46.9	990.9	·	27.7	0.0	116.6	0.0	100.0
4 selected	77.6	43.1	142.7	61.7	19.3	0.0	364.7	·	6.4	0.0	109.2	312.4	100.0
O Oth America	93.0	5.4	40.4	43.2	20.2	118.6	93.9	·	60.2	0.8	128.0	516.6	100.0
West Europe	128.7	416.4	141.3	376.9	179.2	12.9	279.0	13.7	10.2	3.2	16.2	0.5	100.0
EEC(9)		1545.9	530.1	1009.7	262.3	0.0	3.8	2.3	27.2	0.0	16.1	90.5	100.0
N.W.Europe	193.4	97.2	12.8	116.0	97.9	21.8	73.0	0.3	2.6	0.0	3.1	369.5	100.0
S.W.Europe	144.2			315.5	183.8	23.4	7.5	40.0	8.9	0.0	35.5	2.2	100.0
C.W.Europe	195.6	49.6	30.1		115.3	0.0	0.8	0.0	0.0	0.0	0.0	0.0	100.0
W.Eur unalloc	107.6	3.4	185.1	5.3	554.9	0.0	10.5	0.1	0.0	0.0	0.4	0.0	100.0
East Europe	84.4	42.3	163.8	57.3	99.1	347.1	20.0	0.0	0.0	0.0	55.1	0.0	100.0
7 selected	66.9	62.8	196.3	109.0		646.0	0.0	0.0	0.0	0.0	0.4	0.0	100.0
USSR	105.4	20.0	130.2	0.4	211.9		0.1	0.0	0.0	0.0	117.4	0.0	100.0
E.Eur unalloc	0.1	0.7	0.8	1.7		1062.1	0.1	0.0	0.0	0.0	0.0	0.0	100.0
Unallocated	9.2	0.2	0.7	0.2	795.2	316.8		0.2	14.1	0.1	0.9	0.0	100.0
World	100.0	100.0	100.0	100.0	100.0	100.0	100.0	100.0	100.0	100.0	100.0	100.0	100.0

Table II.7. Coefficients of specialization of exporting regions for flat products in 1974

Importers	E.E.C	N.W.E	S.W.E	C.W.E	C.M.E.A	U.R.S.S	N.Amer	L.Amer	Oceania	F.East	Japan	Africa	World
Africa	142.4	20.9	101.5	1.1	5.6	3.0	123.1	0.0	112.3	52.2	112.6	·	100.0
North Africa	241.8	3.7	275.7	0.4	6.9	10.1	12.6	0.0	2.9	0.0	27.0	·	100.0
South Africa	133.9	49.5	97.6	0.6	7.3	0.0	264.7	0.0	220.5	137.3	93.7	·	100.0
Other Africa	111.9	4.6	39.4	1.7	3.7	2.6	52.1	0.1	67.5	4.3	159.6	47.8	100.0
Far East	39.3	10.9	13.9	1.7	32.1	23.4	78.1	0.0	420.8	2.3	212.4	0.0	100.0
3 selected	50.8	13.4	16.3	1.5	56.4	26.3	8.9	0.1	77.7	646.3	208.7	0.0	100.0
Japan	34.6	341.6	0.0	42.3	0.0	0.0	168.5	0.1	12725		0.0	99.3	100.0
Oth Far East	27.1	5.9	11.4	1.6	6.4	20.4	151.5	0.0	696.8	16.9	218.0	1430.0	100.0
Middle East	84.7	3.8	54.6	109.1	79.8	8.8	41.5	0.0	43.7	18.6	153.4	0.0	100.0
2 selected	89.7	5.3	82.1	109.4	125.3	23.3	137.0	0.0	0.0	16.8	127.0	1560.4	100.0
Oth Mid East	84.2	3.7	52.1	109.0	75.6	7.4	32.8	0.0	47.6	466.3	155.8	28.8	100.0
Oceania	26.7	16.8	0.0	1.0	0.0	0.0	101.3	0.0	·	737.2	216.9	0.0	100.0
Oth Oceania	137.1	0.3	0.0	0.7	23.4	0.0	109.0	419.9	129.7	382.7	92.6	9.8	100.0
N. America	121.5	35.4	115.9	12.8	0.0	0.0	·	0.0	1955.1	0.0	127.6	0.0	100.0
N.Am unalloc	255.5	7.2	7.4	15.1	0.0	0.0	399.2	·	19.3	129.5	0.0	20.5	100.0
Oth America	101.4	31.5	25.7	2.5	12.0	23.5	444.8	·	8.0	152.3	106.8	26.3	100.0
4 selected	108.1	39.3	32.3	3.0	11.9	0.0	244.8	131.7	57.7	52.7	96.9	0.7	100.0
O Oth America	78.8	5.2	3.0	0.9	12.4	102.9	92.2	367.2	2.7	27.0	140.1	80.6	100.0
West Europe	131.6	338.6	153.2	279.4	187.3	26.6	170.0	85.2	7.5	46.7	41.8	206.7	100.0
EEC(9)	·	995.3	487.4	750.9	359.8	0.0	25.1	0.6	0.6		45.2	16.6	100.0
N.W.Europe	204.9	·	5.3	92.5	75.8	16.4	102.6	0.0	0.1	39.1	46.0	39.6	100.0
S.W.Europe	157.8	91.0	·	91.2	142.1	68.0	6.9	0.0	0.9	0.3	46.8	10.1	100.0
C.W.Europe	245.6	77.4	32.6	·	79.2	0.0	1.1	0.0	37.2	0.2	10.5	0.0	100.0
W.Eur unalloc	87.6	21.6	246.6	25.0	714.9	0.0	13.7	0.0	0.0	0.1	3.3	0.0	100.0
East Europe	106.1	38.8	189.9	155.0	112.5	509.6	10.9	0.0	0.0	0.1	19.7	0.0	100.0
7 selected	68.4	63.5	73.6	60.1		878.5	17.5	0.0	0.0	0.0	14.2	0.0	100.0
USSR	158.2	4.7	350.5	285.9	268.0	0.0	124.1	0.0	0.0	0.0	27.1	0.0	100.0
E.Eur unalloc	153.2	677.7	900.1	764.4	1057.7	0.0	0.0	1.6	0.8	0.0	0.0	0.0	100.0
Unallocated	1.0	0.2	0.4	0.2		59.4	0.0	0.0	0.0	0.0	0.9	100.0	100.0
World	100.0	100.0	100.0	100.0	100.0	100.0	100.0	100.0	100.0	100.0	100.0	100.0	100.0

Table II.8.   Coefficients of specialization of exporting regions for tubes in 1974

Importers	E.E.C	N.W.E	S.W.E	C.W.E	C.M.E.A	U.R.S.S	N.Amer	L.Amer	Oceania	F.East	Japan	Africa	World
Africa	126.4	65.6	199.4	22.3	2.7	19.5	255.2	88.8	32.2	51.6	70.3	.	100.0
North Africa	127.7	2.8	415.2	18.6	5.2	49.3	203.8	174.0	0.0	49.6	64.8	.	100.0
South Africa	97.1	401.3	35.8	46.4	0.0	0.0	82.7	145.8	411.4	9.7	132.7	.	100.0
Other Africa	133.1	67.9	60.7	22.0	1.3	0.0	316.0	18.0	6.2	58.7	66.2	19.0	100.0
Far East	47.0	29.2	24.4	1.5	20.2	47.2	125.3	1.0	660.0	25.2	197.1	0.0	100.0
3 selected	79.5	39.7	47.8	0.3	41.2	47.8	23.5	0.0	1188.0	1993.9	172.1	548.6	100.0
Japan	21.1	396.3	0.0	14.2	0.0	0.0	376.7	78.4	0.0	0.0	0.0	0.0	100.0
Oth Far East	15.5	9.5	1.9	2.4	0.0	47.8	219.6	0.0	154.1	202.1	226.8	24.3	100.0
Middle East	88.8	7.5	67.0	6.5	8.2	36.8	128.7	37.8	0.0	83.4	140.1	219.5	100.0
2 selected	68.3	6.2	5.7	2.0	0.3	15.9	129.5	0.0	0.0	236.2	182.4	0.0	100.0
Oth Mid East	94.7	7.9	84.7	7.7	10.5	42.8	128.5	48.7	0.0	23.9	127.9	282.6	100.0
Oceania	36.5	107.3	6.7	0.6	0.0	0.0	83.0	8.6	.	197.5	227.6	3.6	100.0
Oth Oceania	108.6	0.0	77.3	0.4	0.0	0.0	93.5	0.0	.	396.1	131.4	0.0	100.0
N. America	43.3	54.7	34.9	10.7	7.1	0.0	.	526.8	53.1	4.7	191.2	136.6	100.0
N.Am unalloc	259.8	27.3	5.3	13.0	0.0	0.0	.	16.8	0.0	1.1	0.0	0.0	100.0
Oth America	60.3	127.3	29.7	19.5	15.7	174.7	541.3	.	5.1	1.9	89.4	9.0	100.0
4 selected	74.7	249.3	53.9	8.4	0.0	0.0	647.7	.	2.4	0.4	61.0	0.0	100.0
O Oth America	47.1	16.5	7.8	29.6	30.0	333.3	444.8	83.2	7.6	28.2	115.2	17.2	100.0
West Europe	121.2	395.3	250.7	510.0	136.8	34.9	66.7	202.0	3.0	68.7	33.0	276.0	100.0
EEC(9)	.	847.4	500.8	894.5	177.9	64.3	125.3	0.0	6.4	0.0	69.8	659.8	100.0
N.W.Europe	212.1	.	48.5	440.1	42.1	15.2	26.6	7.9	1.5	1.3	10.1	0.0	100.0
S.W.Europe	180.8	212.4	.	128.1	198.2	7.2	61.9	0.0	0.0	0.5	12.3	64.3	100.0
C.W.Europe	211.7	136.1	142.1	.	155.1	20.0	1.1	0.0	0.0	2.5	1.0	0.0	100.0
W.Eur unalloc	134.8	225.3	826.8	137.3	217.8	0.0	39.0	3.4	0.0	2.7	1.6	0.0	100.0
East Europe	167.6	47.3	106.0	61.4	208.0	283.8	2.7	3.3	0.0	7.5	20.7	0.0	100.0
7 selected	152.8	62.3	250.6	197.2	.	979.9	7.0	1.7	0.0	0.7	9.7	0.0	100.0
USSR	173.6	41.1	47.0	6.0	292.8	0.0	1.0	3.9	0.0	0.0	25.3	0.0	100.0
E.Eur unalloc	134.9	583.5	608.3	802.3	.	0.0	74.9	0.0	0.0	0.0	0.0	0.0	100.0
Unallocated	8.4	0.6	2.4	0.5	1200.2	53.7	0.1	1.0	7.6	1.4	0.9	0.0	100.0
World	100.0	100.0	100.0	100.0	100.0	100.0	100.0	100.0	100.0	100.0	100.0	100.0	100.0

Table II.9. Coefficients of specialization of exporting regions for all products in 1978

Importers	E.E.C	N.W.E	S.W.E	C.W.E	C.M.E.A	U.R.S.S	N.Amer	L.Amer	Oceania	F.East	Japan	Africa	World
Africa	181.1	48.2	212.7	31.8	21.2	7.2	93.0	133.7	10.9	40.7	68.9	·	100.0
North Africa	190.3	81.3	330.5	53.0	22.1	0.0	37.0	112.1	0.0	11.1	45.9	·	100.0
South Africa	151.9	31.7	37.7	34.3	0.0	0.0	166.7	25.4	46.0	16.5	136.4	·	100.0
Other Africa	172.9	13.1	94.4	8.5	21.5	15.5	149.8	163.5	20.7	74.4	90.2	102.2	100.0
Far East	59.3	24.5	39.9	4.6	19.9	5.9	117.0	66.1	279.0	15.9	210.0		100.0
3 selected	80.3	11.2	34.2	5.5	20.8	0.0	55.8	69.5	194.3		207.3	6.0	100.0
Japan	7.9	224.7	196.3	1.9	158.7	0.0	674.3	503.4	716.7	950.7	0.0	4.0	100.0
Oth Far East	30.8	35.7	41.8	3.3	12.6	14.9	183.5	42.9	384.9		222.8	247.4	100.0
Middle East	121.0	35.2	137.3	12.3	78.6	8.9	88.0	34.2	72.9	168.2	116.8	27.2	100.0
2 selected	102.5	10.7	83.6	13.4	31.8	0.0	224.7	41.6	90.0	220.2	146.9	56.3	100.0
Oth Mid East	126.9	43.0	154.5	12.0	93.6	11.8	44.3	31.9	67.4	151.6	107.1	18.0	100.0
Oceania	36.0	26.0	9.2	7.7	0.0	0.0	99.0	7.4	·	329.6	248.3	12.7	100.0
Oth Oceania	189.4	0.0	0.0	0.0	0.0	0.0	124.4	0.0		15.8	115.8	0.0	100.0
N. America	123.4	80.0	84.0	5.9	20.5	0.0	·	313.0	43.5	222.9	117.7	302.2	100.0
N.Am unalloc	142.0	0.0	0.0	94.7	527.1	0.0		3.3	9.8	0.0	2.6	0.0	100.0
Oth America	107.0	39.2	73.3	12.8	17.5	78.2	630.1	·	150.2	3.5	112.7	110.4	100.0
4 selected	131.1	14.7	91.8	14.6	3.6	0.0	651.8		191.6	4.1	105.7	80.8	100.0
O Oth America	63.3	83.6	39.8	9.6	42.7	220.3	590.6	115.3	75.2	2.4	125.4	164.1	100.0
West Europe	98.6	338.1	164.6	395.0	255.6	30.5	73.9	120.4	70.2	17.1	22.1	88.4	100.0
EEC(9)	·	719.3	281.1	746.4	273.6	21.9	136.3		156.1	33.8	22.8	151.6	100.0
N.W.Europe	224.6	144.5	144.5	161.2	111.9	6.4	13.0	1.1	0.0	8.5	14.5	1.7	100.0
S.W.Europe	139.2	24.3	·	142.2	322.8	70.2	33.8	228.3	0.0	1.0	31.7	75.0	100.0
C.W.Europe	222.4	62.5	122.2	·	169.8	0.5	13.8	0.0	0.0	1.3	10.3	0.2	100.0
W.Eur unalloc	88.3	10.0	176.9	2.5	593.3	67.3	4.4	0.3	2.2	20.8	1.3	0.0	100.0
East Europe	105.8	34.5	82.9	96.0	121.7	502.1	10.7	6.3	0.1	17.6	31.5	0.0	100.0
7 selected	83.7	47.1	88.0	103.1		850.6	26.0	3.0	0.3	0.0	8.5	0.0	100.0
USSR	153.2	37.2	105.6	121.5	238.5	0.0	4.1	10.4	0.0	34.4	56.2	0.0	100.0
E.Eur unalloc	0.0	0.0	0.0	0.0		1385.3	0.0	0.0	0.0	0.0	0.0	0.0	100.0
Unallocated	1.2	0.0	48.3	0.1	549.4	0.0	0.0	0.2	0.0	1595.8	0.0	0.0	100.0
World	100.0	100.0	100.0	100.0	100.0	100.0	100.0	100.0	100.0	100.0	100.0	100.0	100.0

Table II.10. Coefficients of specialization of exporting regions for long products in 1978

Importers	E.E.C	N.W.E	S.W.E	C.W.E	C.M.E.A	U.R.S.S	N.Amer	L.Amer	Oceania	F.East	Japan	Africa	World
Africa	174.0	6.1	204.2	54.9	24.2	..	122.5	155.9	9.2	50.1	8.8	.	100.0
North Africa	163.2	1.5	290.7	93.3	30.2	..	11.3	78.9	0.0	21.3	6.2	.	100.0
South Africa	194.0	110.9	10.6	161.6	0.0	..	257.6	13.4	543.2	0.0	19.2	.	100.0
Other Africa	187.8	10.3	95.5	3.3	16.7	..	264.1	257.8	12.4	88.0	11.9	147.4	100.0
Far East	64.2	30.9	17.8	18.6	14.6	..	28.4	20.0	384.3	29.6	234.2	2.9	100.0
3 selected	81.5	14.2	22.1	21.3	17.0	..	12.7	31.6	395.1	.	223.3	0.0	100.0
Japan	44.5	40.6	0.0	10.5	0.0	..	298.3	0.0	5.0	3729.9	0.0	.	100.0
Oth Far East	34.4	59.8	10.7	14.0	10.6	..	50.0	0.2	373.7	.	258.5	402.9	100.0
Middle East	122.8	19.6	121.7	4.5	83.8	..	51.5	35.9	4.6	346.0	83.1	10.9	100.0
2 selected	108.9	13.8	79.1	7.2	39.0	..	123.5	64.3	18.7	495.2	121.5	22.2	100.0
Oth Mid East	127.3	21.5	135.6	3.5	98.5	..	28.0	26.6	0.0	297.2	70.5	7.3	100.0
Oceania	55.7	50.2	2.1	52.7	0.0	..	123.4	0.0	.	54.8	279.0	2.8	100.0
Oth Oceania	175.8	0.0	0.0	0.0	0.0	..	66.5	0.0	.	0.8	123.6	0.0	100.0
N. America	104.9	39.6	36.0	8.7	18.6	..	.	437.5	59.9	55.8	138.4	285.6	100.0
N.Am unalloc	114.2	0.0	0.0	0.0	517.1	..	1151.5	2.4	22.8	0.0	0.0	0.0	100.0
Oth America	98.2	20.8	101.2	62.9	26.8	..	1422.1	.	29.8	1.8	76.2	176.5	100.0
4 selected	95.0	22.0	121.1	88.5	2.6	..	699.6	.	0.1	0.0	74.7	31.8	100.0
O Oth America	103.5	18.7	68.0	20.1	67.2	..	72.2	61.6	79.3	4.7	78.8	418.2	100.0
West Europe	93.3	384.6	142.4	398.6	185.0	..	123.3	129.1	17.5	4.0	10.0	78.1	100.0
EEC(9)		791.0	261.7	776.9	166.9	..	9.4	0.0	36.8	7.7	11.8	111.1	100.0
N.W.Europe	196.5	26.7	63.2	61.8	150.8	..	49.2	0.0	0.0	0.0	2.6	0.0	100.0
S.W.Europe	141.5	25.0		113.1	288.5	..	16.9	1.9	0.4	1.3	19.7	129.5	100.0
C.W.Europe	219.0		47.9		96.7	..	0.7	0.0	0.0	0.0	1.5	0.0	100.0
W.Eur unalloc	97.5	4.2	66.8	2.4	430.0	..	43.3	0.1	0.0	4.6	0.0	0.0	100.0
East Europe	94.5	64.7	171.8	78.3	251.3	..	182.2	0.0	0.0	138.3	46.8	0.0	100.0
7 selected	161.4	61.5	202.9	313.2		..	0.1	0.0	0.0	0.3	22.7	0.0	100.0
USSR	73.7	65.6	162.1	5.3	329.4	..	0.0	0.0	0.0	181.2	54.3	0.0	100.0
E.Eur unalloc	.	.	.	.	863.7	..	0.0	0.2	0.1	0.0	0.0	.	100.0
Unallocated	2.9	0.0	48.7	0.0	100.0	..	0.0	0.2	0.1	0.0	0.0	0.0	100.0
World	100.0	100.0	100.0	100.0	100.0	..	100.0	100.0	100.0	100.0	100.0	100.0	100.0

Table II.11. Coefficients of specialization of exporting regions for flat products in 1978

Importers	E.E.C	N.W.E	S.W.E	C.W.E	C.M.E.A	U.R.S.S	N.Amer	L.Amer	Oceania	F.East	Japan	Africa	World
Africa	151.2	2.4	65.3	8.2	8.7	...	55.6	0.5	14.3	9.6	110.4	.	100.0
North Africa	241.9	2.9	160.2	29.7	24.5	...	107.4	0.0	0.0	0.8	12.0	.	100.0
South Africa	127.9	11.8	0.0	2.6	0.0	...	125.0	11.7	33.5	55.6	131.2	.	100.0
Other Africa	122.1	1.7	37.4	1.3	3.9	...	33.7	0.0	17.9	9.7	142.3	.	100.0
Far East	48.5	15.5	12.9	0.4	21.6	...	156.0	29.1	263.4	21.4	186.1	52.3	100.0
3 selected	67.6	8.7	2.7	0.4	26.8	...	39.3	0.0	278.1	.	179.7	0.3	100.0
Japan	6.9	70.9	82.3	3.3	68.6	...	729.0	981.0	2.1	3011.8	0.0	20.7	100.0
Oth Far East	23.4	23.6	25.6	0.3	13.6	...	304.4	52.6	247.9	.	197.9	123.1	100.0
Middle East	96.9	13.1	55.0	24.3	101.1	...	79.9	0.0	7.6	20.0	134.0	11.8	100.0
2 selected	87.1	19.9	35.3	41.2	69.3	...	445.0	0.0	41.4	29.7	123.9	47.2	100.0
Oth Mid East	98.3	12.2	57.8	21.9	105.6	...	27.9	0.0	2.8	18.6	135.5	6.7	100.0
Oceania	30.6	14.9	16.8	0.8	0.0	...	81.8	0.0	.	496.5	200.1	18.9	100.0
Oth Oceania	174.8	0.0	0.0	0.0	0.0	...	190.5	0.0	.	6.5	89.4	0.0	100.0
N. America	112.2	101.3	164.1	4.3	28.2	...	.	299.0	120.9	359.2	87.3	341.9	100.0
N.Am unalloc	128.5	0.0	0.0	0.0	618.3	...	.	0.0	0.0	0.0	0.0	0.0	100.0
Oth America	92.4	9.9	123.6	4.2	23.2	...	394.2	.	1.6	10.6	131.2	79.7	100.0
4 selected	108.8	13.2	176.5	2.5	0.2	...	475.2	.	1.7	14.9	109.9	126.2	100.0
O Oth America	66.6	4.7	40.5	6.9	59.4	...	266.9	146.0	1.5	3.8	164.8	6.4	100.0
West Europe	104.4	316.4	176.0	329.1	233.4	...	95.2	296.9	20.3	43.6	25.7	44.5	100.0
EEC(9)	.	706.2	191.2	659.3	279.6	...	187.5	.	47.8	101.8	31.5	92.9	100.0
N.W.Europe	194.9	.	266.5	120.1	125.7	...	8.4	.	0.1	0.0	20.7	3.0	100.0
S.W.Europe	161.1	23.0	.	118.5	294.2	...	50.3	85.5	0.0	0.0	24.2	19.4	100.0
C.W.Europe	203.3	91.4	311.0	.	95.6	...	16.6	0.0	0.0	4.9	18.6	0.4	100.0
W.Eur unalloc	55.6	18.6	245.2	1.8	788.7	...	0.0	0.3	15.4	0.0	4.2	0.0	100.0
East Europe	198.7	52.5	74.1	237.7	151.4	...	15.1	0.0	0.8	19.0	12.2	0.0	100.0
7 selected	209.5	157.5	140.2	220.8	.	...	49.3	0.0	4.4	0.0	22.8	0.0	100.0
USSR	196.1	27.3	58.2	241.7	187.8	...	6.9	0.0	0.0	23.6	9.6	0.0	100.0
E.Eur unalloc	.	.	.	.	.	...	.	.	.	.	.	.	.
Unallocated	1.3	0.0	154.5	0.0	1061.2	...	0.0	0.3	0.1	0.1	0.0	0.0	100.0
World	100.0	100.0	100.0	100.0	100.0	100.0	100.0	100.0	100.0	100.0	100.0	100.0	100.0

Table II.12. Coefficients of specialization of exporting regions for tubes in 1978

Importers	E.E.C	N.W.E	S.W.E	C.W.E	C.M.E.A	U.R.S.S	N.Amer	L.Amer	Oceania	F.East	Japan	Africa	World
Africa	125.3	42.1	104.8	46.5	5.5	...	110.9	165.2	6.7	42.3	101.0	.	100.0
North Africa	110.2	6.3	127.0	37.7	3.3	...	53.8	236.7	0.0	11.1	119.9	.	100.0
South Africa	117.5	27.1	102.4	21.2	0.0	...	149.6	10.3	114.6	0.0	118.4	.	100.0
Other Africa	160.6	124.9	55.6	71.2	11.4	...	231.1	35.5	0.6				100.0
Far East	84.3	53.0	96.5	1.1	19.0	...	41.6	69.2	502.4	120.4	55.4		100.0
3 selected	111.3	43.2	136.9	0.2	26.7	...	7.0	98.3	667.6	12.0	147.2	53.1	100.0
Japan	4.7	873.8	0.0	0.0	0.0	...	419.9	0.0	101.7		118.3	0.0	100.0
Oth Far East	20.2	57.3	0.2	3.5	0.6	...	116.8	0.0	109.3	1752.7	0.0	0.0	100.0
Middle East	76.9	6.1	25.6	19.1	15.9	...	197.7	26.3	0.0	131.7	221.1	183.7	100.0
2 selected	72.4	4.9	8.8	12.0	1.3	...	212.5	0.0	0.0	93.3	144.3	0.0	100.0
Oth Mid East	83.5	7.7	46.9	28.4	17.8	...	181.5	59.6		181.4	158.5	0.0	100.0
Oceania	29.4	81.8	10.7	4.8	0.0	...	104.4	35.6		440.6	128.2	0.0	100.0
Oth Oceania	139.7	0.0	0.0	0.0	0.0	...	69.1	0.0		46.9	165.9	0.0	100.0
N. America	36.2	45.4	104.9	1.4	22.6	...			151.8	401.5	111.7	0.0	100.0
N.Am unalloc	240.5	0.0	0.0	0.0	0.0	...	496.9	377.2	0.0	3.6	142.9	397.3	100.0
Oth America	133.5	28.7	17.4	6.7	13.9	...	454.9	9.8	35.3	0.8	25.9	0.0	100.0
4 selected	152.6	29.7	18.4	1.4	0.0	...	701.1		0.0	0.9	87.5	5.5	100.0
O Oth America	40.8	23.9	12.8	32.4	81.6	...				0.3	75.3	0.0	100.0
West Europe	119.7	511.8	311.3	763.4	156.2	...	78.0	14.0	206.7	23.9	146.9	32.1	100.0
EEC(9)		1045.7	641.6	1421.1	215.5	...	135.8	25.2	0.8	28.5	11.1	179.5	100.0
N.W.Europe	211.6		50.0	519.4	92.4	...	25.7	6.5	1.8	37.2	21.2	367.9	100.0
S.W.Europe	173.1	167.2		151.6	211.8	...	102.4	12.7	0.0	0.0	1.8	0.0	100.0
C.W.Europe	238.8	203.6	46.0		78.1	...	3.2	0.0	0.0	0.0	12.6	177.7	100.0
W.Eur unalloc	102.7	50.6	1132.5	6.6	72.6	...	53.1	1.9	0.0	212.0	0.2	0.0	100.0
East Europe	139.0	80.2	81.3	40.0	209.2	...	2.6	24.3	0.0	1.4	1.0	0.0	100.0
7 selected	199.9	247.1	266.5	168.4		...	4.2	26.4	0.0	0.1	70.2	0.0	100.0
USSR	128.7	52.2	50.3	18.5	244.3	...	2.3	23.9	0.0	1.6	18.9	0.0	100.0
E.Eur unalloc	0.0	0.0	0.0	0.0		...	0.0	6663.0	0.0	0.0	78.8	0.0	100.0
Unallocated	2.0	0.0	51.3	0.9	1381.1	...	0.0	0.5	0.9	0.1	0.0	0.0	100.0
World	100.0	100.0	100.0	100.0	100.0		100.0	100.0	100.0	100.0	100.0	100.0	100.0

Table II.13. Coefficients of specialization of exporting regions for all products in 1982

Importers	E.E.C	N.W.E	S.W.E	C.W.E	C.M.E.A	U.R.S.S	N.Amer	L.Amer	Oceania	F.East	Japan	Africa	World
Africa	178.9	32.1	384.9	22.4	12.7	1.6	97.5	121.0	3.4	22.8	63.2	·	100.0
North Africa	185.7	42.8	638.2	32.6	3.4	0.0	48.6	100.8	0.0	7.4	15.3	·	100.0
South Africa	161.5	95.8	19.3	34.4	24.6	0.0	68.0	21.8	13.8	0.0	172.7	·	100.0
Other Africa	172.2	14.8	112.0	9.3	20.3	3.5	157.5	152.3	6.6	42.7	111.8	42.4	100.0
Far East	43.7	17.4	17.0	2.1	34.2	17.5	123.4	104.1	349.0	129.7	217.9	0.0	100.0
3 selected	69.3	25.1	32.6	4.2	14.4	0.0	114.9	52.4	220.5	1275.2	234.8	0.0	100.0
Japan	1.6	5.2	0.0	0.2	·	0.0	25.1	361.3	1.1	·	0.0	80.8	100.0
Oth Far East	33.7	14.2	9.2	1.0	81.0	33.3	148.4	91.0	508.1	107.7	248.0	0.1	100.0
Middle East	71.2	29.8	223.9	29.3	56.2	7.0	150.7	119.3	26.3	196.2	143.2	0.0	100.0
2 selected	54.0	34.2	72.7	19.5	100.2	0.0	126.7	127.5	28.8	39.0	182.4	0.1	100.0
Oth Mid East	84.5	26.4	341.2	36.8	0.1	12.4	169.3	112.9	24.4	123.7	112.7	0.3	100.0
Oceania	40.6	18.3	1.7	8.4	2.4	0.0	41.8	57.2		4.5	266.7	0.0	100.0
Oth Oceania	217.0	0.0	11.5	6.1	5.9	0.0	86.5	0.1	108.6	106.4	136.7	370.3	100.0
N. America	145.1	106.3	66.7	20.1	0.0	0.0		255.2	0.0	0.1	113.9	0.0	100.0
N.Am unalloc	377.4	8.3	4.2	0.0	29.3	112.3	508.7	0.7	183.1	29.8	3.8	117.4	100.0
Oth America	124.3	21.2	73.4	22.4	11.1	0.0	531.2		279.7	9.3	102.3	48.4	100.0
4 selected	158.4	31.5	81.6	13.9	52.9	258.1	479.6	107.4	57.7	56.3	106.3	207.0	100.0
O Oth America	80.0	7.9	62.7	33.3	189.0	29.6	110.4	206.6	16.8	52.5	97.0	180.5	100.0
West Europe	135.8	338.1	115.8	326.3	247.8	20.9	160.2	18.8	30.6	23.4	14.0	308.7	100.0
EEC(9)		711.3	206.9	669.3	94.1	11.0	13.8	43.2	0.1	227.6	10.0	0.2	100.0
N.W.Europe	223.7		83.6	98.0	180.4	67.6	132.4	17.0	13.6	6.0	24.0	169.5	100.0
S.W.Europe	224.3	94.1		82.8	128.7	0.7	29.0	302.0	0.0	1.0	18.9	20.8	100.0
C.W.Europe	293.0	70.2	100.0		353.8	41.6	17.8	3.1	0.0	1.9	2.7	0.0	100.0
W.Eur unalloc	160.6	24.3	123.8	5.0	151.5	480.8	0.5	0.3	0.0	1.3	4.1	0.0	100.0
East Europe	103.2	21.7	46.9	120.2	259.2	1069.6	0.6	5.1	0.0	0.0	48.7	0.0	100.0
7 selected	37.2	29.6	81.5	56.5		0.0	0.7	0.0	0.0	2.2	3.2	0.0	100.0
USSR	159.9	23.9	43.7	180.2	417.1	1308.3	0.0	0.0	0.0	0.0	81.9	0.0	100.0
E.Eur unalloc	0.0	0.0	0.0	0.0		0.0	0.0	0.0	0.0	843.5	0.0	0.0	100.0
Unallocated	0.7	0.0	0.4	0.0		0.0	0.0	0.0	59.6	0.0	0.0	0.0	100.0
World	100.0	100.0	100.0	100.0	100.0	100.0	100.0	100.0	100.0	100.0	100.0	100.0	100.0

Table II.14. Coefficients of specialization of exporting regions for long products in 1982

Importers	E.E.C	N.W.E	S.W.E	C.W.E	C.M.E.A	U.R.S.S	N.Amer	L.Amer	Oceania	F.East	Japan	Africa	World
Africa	196.4	19.1	306.0	37.5	3.8	...	63.8	128.6	0.5	24.8	12.7	·	100.0
North Africa	152.8	13.5	405.7	42.2	0.4	...	33.6	102.0	0.0	13.9	5.5	·	100.0
South Africa	325.4	98.3	8.5	158.7	0.0	...	112.8	42.2	36.7	0.0	51.5	·	100.0
Other Africa	283.0	28.7	108.3	24.5	11.1	...	124.8	185.7	0.5	48.1	26.5	57.3	100.0
Far East	61.7	17.5	8.0	8.3	6.1	...	109.5	29.2	277.4	38.0	243.4	0.0	100.0
3 selected	80.4	36.4	17.2	20.2	9.6	...	219.9	72.4	225.1	·	218.2	0.0	100.0
Japan	20.4	36.0	0.0	0.0	0.0	...	148.5	48.7	8.9	2633.3	0.0	0.0	100.0
Oth Far East	52.8	7.2	3.3	2.3	4.4	...	50.8	6.2	310.7	·	262.0	88.6	100.0
Middle East	43.8	19.4	122.8	16.0	60.9	...	80.8	140.3	58.6	257.9	146.7	0.0	100.0
2 selected	33.4	9.2	40.1	8.2	47.7	...	143.9	161.2	81.5	471.7	171.8	0.0	100.0
Oth Mid East	55.5	30.8	215.0	24.7	75.7	...	10.4	117.0	33.1	19.1	118.6	0.0	100.0
Oceania	87.4	27.3	0.6	54.7	0.0	...	72.5	3.9	·	21.1	241.7	0.1	100.0
Oth Oceania	246.8	0.0	0.0	13.2	0.0	...	156.3	0.1	0.1	6.8	134.6	0.0	100.0
N. America	200.4	41.7	51.5	9.8	0.1	...	·	304.5	97.2	71.1	91.2	289.3	100.0
N.Am unalloc	421.3	29.3	0.0	0.0	0.0	...	·	1.8	0.0	0.0	21.1	0.0	100.0
Oth America	162.4	21.1	40.6	107.5	59.1	...	1312.1	·	336.4	46.9	46.9	225.3	100.0
4 selected	179.6	20.9	27.2	87.6	59.2	...	1362.7	·	0.0	1.6	50.7	228.2	100.0
O Oth America	148.1	21.2	51.7	124.1	58.9	...	1269.9	·	616.3	84.6	43.7	222.8	100.0
West Europe	131.4	378.9	103.5	380.4	148.8	...	70.4	48.5	88.4	4.9	7.2	254.2	100.0
EEC(9)	·	681.7	136.0	688.1	164.1	...	116.1	65.9	169.5	8.2	8.1	410.7	100.0
N.W.Europe	265.6	·	130.5	39.2	121.6	...	4.7	1.6	0.0	2.8	2.9	0.0	100.0
S.W.Europe	236.7	118.8	·	99.3	167.8	...	54.2	44.5	0.0	1.0	15.8	250.6	100.0
C.W.Europe	324.4	25.8	78.9	·	108.5	...	3.1	0.0	0.0	0.0	0.1	0.5	100.0
W.Eur unalloc	195.8	22.4	76.6	8.8	106.8	...	2.8	622.8	0.0	0.9	0.0	0.0	100.0
East Europe	72.9	55.7	91.5	71.1	376.4	...	0.1	0.0	0.0	0.0	30.4	0.0	100.0
7 selected	155.2	94.4	328.4	556.2		...	0.7	0.0	0.0	0.0	2.1	0.0	100.0
USSR	62.2	50.6	60.5	7.6	425.6	...	0.0	0.0	0.0	0.0	34.0	0.0	100.0
E.Eur unalloc	·	·				...							
Unallocated	1.1	0.0	0.7	0.0	609.3	...	0.0	0.1	0.2	190.4	0.0	0.0	100.0
World	100.0	100.0	100.0	100.0	100.0	...	100.0	100.0	100.0	100.0	100.0	100.0	100.0

Table II.15. Coefficients of specialization of exporting regions for flat products in 1982

Importers	E.E.C	N.W.E	S.W.E	C.W.E	C.M.E.A	U.R.S.S	N.Amer	L.Amer	Oceania	F.East	Japan	Africa	World
Africa	166.6	15.4	99.4	5.4	4.7	...	58.6	31.9	11.9	57.9	129.0	•	100.0
North Africa	305.6	44.0	351.2	13.6	0.1	...	54.0	13.7	0.0	5.0	2.3	•	100.0
South Africa	138.2	32.3	0.0	18.2	0.0	...	21.4	11.9	15.3	0.0	177.7	•	100.0
Other Africa	118.0	3.8	14.9	1.5	6.6	...	62.9	40.0	16.0	81.3	171.6	29.4	100.0
Far East	39.6	14.3	34.3	0.3	30.9	...	104.7	111.2	215.0	137.0	190.6	0.0	100.0
3 selected	68.5	19.5	57.1	0.5	47.2	...	79.7	31.6	112.7	•	203.5	0.0	100.0
Japan	0.5	3.1	•	0.2	0.0	...	20.8	436.5	0.4	1339.1	0.0	0.0	100.0
Oth Far East	19.5	11.7	19.1	0.1	21.6	...	149.1	117.1	366.9	•	221.1	65.7	100.0
Middle East	86.0	9.1	316.7	39.0	91.1	...	233.0	63.3	57.2	36.6	115.7	0.0	100.0
2 selected	66.9	7.2	99.2	71.0	47.7	...	74.1	0.0	38.2	92.7	185.6	0.0	100.0
Oth Mid East	92.5	9.7	390.3	28.2	105.7	...	286.8	84.7	63.6	17.7	92.1	0.0	100.0
Oceania	22.2	12.9	0.8	0.4	0.0	...	19.9	32.8	•	180.8	244.5	0.4	100.0
Oth Oceania	201.0	0.0	0.0	0.0	0.0	...	7.2	0.0	•	3.9	135.8	0.0	100.0
N. America	122.2	119.0	119.3	2.5	6.1	...	•	228.1	262.9	94.9	98.7	468.5	100.0
N.Am unalloc	366.8	0.0	0.0	0.0	0.0	...	•	0.0	0.0	0.0	0.0	•	100.0
Oth America	103.2	14.0	144.3	6.4	35.7	...	300.9	•	9.3	72.2	127.2	169.3	100.0
4 selected	133.1	23.7	161.2	4.6	0.3	...	408.0	•	3.8	34.2	106.8	29.4	100.0
O Oth America	71.8	4.0	126.5	8.3	72.8	...	188.5	123.0	15.1	112.1	148.6	316.0	100.0
West Europe	120.2	283.9	99.6	230.0	176.1	...	123.8	123.0	0.1	120.5	9.8	91.2	100.0
EEC(9)	•	623.2	180.1	443.6	241.3	...	169.4	234.2	0.2	6.0	8.8	159.0	100.0
N.W.Europe	198.3	46.7	61.3	69.9	73.4	...	12.5	21.9	0.0	502.7	5.7	0.2	100.0
S.W.Europe	177.3		73.8	114.0	194.5	...	200.4	69.1	0.0	4.8	18.5	89.2	100.0
C.W.Europe	291.0	73.8		97.6	97.6	...	39.1	26.4	0.0	0.0	4.9	33.1	100.0
W.Eur unalloc	172.7	23.5	35.2	3.5	456.0	...	9.1	1.1	0.0	0.6	5.6	0.0	100.0
East Europe	187.8	34.8	90.3	297.1	153.8	...	1.0	1.5	0.0	3.1	47.0	0.0	100.0
7 selected	245.3	206.2	622.4	117.6	•	...	5.4	0.0	0.0	6.1	6.1	0.0	100.0
USSR	183.4	21.6	49.3	310.9	165.7	...	0.7	1.7	0.0	3.3	50.1	0.0	100.0
E.Eur unalloc						...							
Unallocated	1.2	0.0	0.5	0.0	797.5	...	0.0	0.0	0.0	258.0	0.0	0.0	100.0
World	100.0	100.0	100.0	100.0	100.0	...	100.0	100.0	100.0	100.0	100.0	100.0	100.0

Table II.16. Coefficients of specialization of exporting regions for tubes in 1982

Importers	E.E.C	N.W.E	S.W.E	C.W.E	C.M.E.A	U.R.S.S	N.Amer	L.Amer	Oceania	F.East	Japan	Africa	World
Africa	164.4	55.8	226.9	37.7	10.9	...	120.9	156.7	8.3	37.0	54.8	.	100.0
North Africa	169.6	18.7	333.4	66.6	14.3	...	48.8	292.6	0.0	28.5	35.9	.	100.0
South Africa	85.0	321.0	52.3	8.4	0.0	...	115.0	33.5	97.9	0.0	148.7	.	100.0
Other Africa	174.4	46.6	131.2	8.1	8.8	...	211.6	13.9	0.0	55.1	58.7	197.2	100.0
Far East	39.7	34.7	52.7	2.7	6.8	...	110.2	68.4	43.9	29.8	190.6	0.0	100.0
3 selected	60.2	49.1	93.9	1.9	13.4	...	97.9	114.7	61.8	.	170.4	2.7	100.0
Japan	27.9	199.9	0.7	3.4	0.0	...	333.6	0.3	0.0	1692.2	0.0	.	100.0
Oth Far East	26.6	20.2	27.1	3.3	2.7	...	111.7	39.8	33.4	.	209.5	333.2	100.0
Middle East	95.7	12.8	223.2	38.2	35.6	...	263.9	76.0	1.1	188.3	90.3	1.2	100.0
2 selected	99.7	5.4	19.0	15.1	1.3	...	408.5	41.4	0.0	213.8	108.6	0.0	100.0
Oth Mid East	91.9	19.7	412.7	59.7	67.5	...	129.7	108.2	2.0	164.5	73.4	2.4	100.0
Oceania	41.1	42.4	5.8	0.7	0.4	...	87.3	159.3	.	165.4	178.2	0.8	100.0
Oth Oceania	207.5	0.0	128.2	0.0	0.0	...	334.1	0.0	0.3	18.4	33.0	0.0	100.0
N. America	80.3	50.4	81.3	59.1	17.5	...	.	297.3	0.0	259.1	112.3	287.3	100.0
N.Am unalloc	282.9	9.2	7.4	0.0	0.0	...	557.4	0.6	1.3	0.1	0.0	0.0	100.0
Oth America	119.1	33.4	62.4	8.9	9.7	...	508.2	.	.	18.8	101.8	14.5	100.0
4 selected	123.4	36.0	74.8	7.0	1.0	...	692.1	.	4.7	0.1	103.2	0.0	100.0
O Oth America	107.2	26.1	28.5	14.2	33.5	...	59.2	7.1	2.5	70.3	97.9	54.3	100.0
West Europe	107.0	651.6	158.1	621.8	136.6	...	122.0	14.1	5.6	8.2	43.7	110.3	100.0
EEC(9)	.	1493.6	329.8	1365.3	184.6	...	13.2	3.3	0.9	20.2	23.6	274.6	100.0
N.W.Europe	148.1	.	53.4	220.5	98.4	...	13.2	3.3	0.0	0.0	82.0	1.2	100.0
S.W.Europe	178.8	209.1	.	72.6	140.9	...	39.9	3.3	0.0	1.1	49.4	15.8	100.0
C.W.Europe	243.6	286.4	64.4	0.0	94.8	...	5.2	0.1	0.0	0.0	1.2	0.0	100.0
W.Eur unalloc	217.2	53.3	160.6	0.0	66.0	...	173.7	2.8	0.0	22.9	16.2	0.0	100.0
East Europe	134.6	22.3	51.3	40.7	205.4	...	0.7	14.2	0.0	1.9	86.6	0.0	100.0
7 selected	144.8	165.3	521.8	256.1	.	...	0.0	11.3	0.0	0.0	43.2	0.0	100.0
USSR	134.2	15.9	30.4	31.2	214.5	...	0.7	14.3	0.0	2.0	88.6	0.0	100.0
E.Eur unalloc	.	.	.	.	.	...	.	.	.	323.8	0.0	0.0	100.0
Unallocated	2.0	0.0	0.3	0.0	1109.9	...	0.0	0.2	5420.9	.	.	.	100.0
World	100.0	100.0	100.0	100.0	100.0	100.0	100.0	100.0	100.0	100.0	100.0	100.0	100.0

Table II.17. Changes in the geographic orientation of exports of steel products including intra-regional trade

	Economic European community				North European countries				South European countries				Central European countries			
Years	1983		1984		1983		1984		1983		1984		1983		1984	
Importers	1000t	%	1000t	%	1000t	%	1000t	%	1000t	%	1000t	%	1000t	%	1000t	%
Africa	2063	3.9	2267	3.8	42	1.0	46	1.0	1367	17.2	1227	12.5	45	1.3	71	1.7
Algeria	485	0.9	603	1.0	19	0.5	20	0.4	532	6.7	327	3.3	39	1.1	64	1.5
Liby.Arab Jam.	236	0.4	246	0.4	3	0.1	2	0.0	266	3.3	34	0.3	2	0.0	4	0.1
Morocco	215	0.4	340	0.6	0	0.0	0	0.0	310	3.9	327	3.3	0	0.0	0	0.0
Tunisia	198	0.4	194	0.3	0	0.0	0	0.0	90	1.1	115	1.2	1	0.0	0	0.0
North Africa	1135	2.2	1383	2.3	23	0.5	22	0.5	1197	15.0	803	8.2	41	1.2	68	1.6
South Africa	73	0.1	90	0.2	4	0.1	8	0.2	2	0.0	3	0.0	2	0.1	1	0.0
Other Africa	855	1.6	794	1.3	15	0.3	16	0.3	169	2.1	421	4.3	2	0.1	2	0.0
Far East	3284	6.2	3582	6.0	103	2.4	250	5.2	566	7.1	1069	10.9	12	0.3	20	0.5
China	1220	2.3	1256	2.1	5	0.1	133	2.8	157	2.0	447	4.6	1	0.0	13	0.3
India	502	1.0	929	1.6	20	0.5	20	0.4	54	0.7	58	0.6	4	0.1	3	0.1
Korea,Rep.of	60	0.1	133	0.2	4	0.1	7	0.2	12	0.1	10	0.1	0	0.0	0	0.0
3 of Far East	1782	3.4	2318	3.9	28	0.7	160	3.3	223	2.8	515	5.2	5	0.1	16	0.4
Japan	7	0.0	28	0.0	5	0.1	8	0.2	78	1.0	296	3.0	2	0.0	0	0.0
Other Far East	1495	2.8	1236	2.1	70	1.6	63	1.7	265	3.3	258	2.6	5	0.1	3	0.1
Middle East	2840	5.4	2805	4.7	129	3.0	49	1.0	2571	32.2	2544	25.9	131	3.6	104	2.5
Egypt	338	0.6	292	0.5	7	0.2	8	0.2	250	3.1	535	5.4	10	0.3	18	0.4
Saudi Arabia	512	1.0	554	0.9	51	1.2	22	0.5	164	2.1	267	2.7	2	0.0	4	0.1
2 of M.East	850	1.6	846	1.4	58	1.4	30	0.6	414	5.2	802	8.2	12	0.3	22	0.5
Other M.East	1990	3.8	1959	3.3	71	1.7	19	0.4	2157	27.1	1742	17.7	119	3.3	82	2.0
Oceania	126	0.2	116	0.2	6	0.1	11	0.2	1	0.0	13	0.1	2	0.1	3	0.1
Australia	72	0.1	48	0.1	5	0.1	9	0.2	1	0.0	12	0.1	2	0.1	3	0.1
New Zealand	33	0.1	38	0.1	1	0.0	1	0.0	0	0.0	0	0.0	0	0.0	0	0.0
Other Oceania	22	0.0	30	0.1	0	0.0	0	0.0	0	0.0	1	0.0	0	0.0	0	0.0
North America	4479	8.5	6782	11.4	472	11.1	971	20.2	706	8.9	1633	16.6	28	0.8	151	3.7
Canada	373	0.7	667	1.1	21	0.5	68	1.4	26	0.3	114	1.2	5	0.2	17	0.4
U.S.of America	4102	7.8	6019	10.2	450	10.6	903	18.8	681	8.5	1518	15.5	23	0.6	135	3.3
N.America una.	5	0.0	96	0.2	-	-	-	-	0	0.0	1	0.0	-	-	-	-
Other America	1116	2.1	1220	2.1	17	0.4	39	0.8	113	1.4	176	1.8	11	0.3	14	0.3
Argentina	328	0.6	220	0.4	6	0.1	8	0.2	14	0.2	4	0.0	1	0.0	1	0.0
Brazil	48	0.1	57	0.1	3	0.1	3	0.1	1	0.0	-	-	6	0.2	1	0.0
Mexico	184	0.3	367	0.6	5	0.1	6	0.1	11	0.1	29	0.3	1	0.0	2	0.0
Venezuela	92	0.2	74	0.1	1	0.0	1	0.0	4	0.1	7	0.1	0	0.0	1	0.0
4 of O.America	652	1.2	718	1.2	14	0.3	18	0.4	31	0.4	39	0.4	7	0.2	4	0.1
Oth.Oth.America	465	0.9	502	0.8	4	0.1	21	0.4	82	1.0	137	1.4	3	0.1	10	0.2
Western Europe	33400	63.4	36975	62.4	3341	78.3	3301	68.8	1819	22.8	2221	22.6	2406	67.0	2628	53.7
B.L.E.U.	2792	5.3	3116	5.3	121	2.6	103	2.1	124	1.6	115	1.2	32	0.9	39	1.0
Denmark	805	1.5	850	1.4	426	10.0	473	9.8	9	0.1	6	0.1	48	1.4	50	1.2
France	6317	12.0	6994	11.8	181	4.2	156	3.2	250	3.1	227	2.3	135	3.8	144	3.5
Germany,Fed.Rep	6862	13.0	6740	11.4	861	20.2	804	16.8	412	5.2	518	5.3	1306	36.4	1349	32.7
Ireland	303	0.6	284	0.5	15	0.3	31	0.6	13	0.2	18	0.2	5	0.1	3	0.1
Italy	3320	6.3	4191	7.1	96	2.3	126	2.6	211	2.6	260	2.7	372	10.4	475	11.5
Netherlands	2727	5.2	3130	5.3	294	6.9	266	5.6	62	0.8	59	0.6	81	2.3	101	2.5
United Kingdom	2387	4.5	2673	4.5	403	9.4	412	8.6	172	2.2	199	2.0	73	2.0	83	2.0
EEC(9)	25513	48.4	27980	47.2	2396	56.2	2371	49.4	1254	15.7	1401	14.3	2053	57.2	2244	54.4
Finland	249	0.5	280	0.5	158	3.7	146	3.0	51	0.6	38	0.4	12	0.3	13	0.3
Iceland	16	0.0	19	0.0	15	0.3	18	0.4	11	0.1	4	0.0	0	0.0	0	0.0
Norway	947	1.8	848	1.4	276	6.5	287	6.0	35	0.4	31	0.3	12	0.3	15	0.4
Sweden	1301	2.5	1414	2.4	268	6.3	279	5.8	85	1.1	43	0.4	64	1.8	68	1.6
Northern Europe	2513	4.8	2561	4.3	717	16.8	730	15.2	183	2.3	117	1.2	89	2.5	96	2.3
Greece	1015	1.9	1164	2.0	45	1.1	44	0.9	16	0.2	28	0.3	3	0.1	5	0.1
Portugal	357	0.7	414	0.7	11	0.3	14	0.3	48	0.6	69	0.7	3	0.1	2	0.1
Spain	1003	1.9	1218	2.1	40	0.9	19	0.4	3	0.0	1	0.0	9	0.3	11	0.3
Turkey	359	0.7	722	1.2	38	0.9	6	0.1	144	1.8	448	4.6	7	0.2	12	0.3
Yugoslavia	280	0.5	363	0.6	13	0.3	14	0.3	49	0.6	31	0.3	80	2.2	88	2.1
Southern Europe	3014	5.7	3880	6.5	147	3.4	97	2.0	261	3.3	577	5.9	102	2.8	119	2.9
Austria	675	1.3	749	1.3	24	0.6	22	0.5	42	0.5	56	0.6	56	1.6	50	1.2
Switzerland	1624	3.1	1733	2.9	39	0.9	33	0.7	66	0.8	50	0.5	106	3.0	118	2.9
Central Europe	2299	4.4	2481	4.2	63	1.5	55	1.1	108	1.4	106	1.1	162	4.5	168	4.1
W.Europe unallo	61	0.1	72	0.1	18	0.4	47	1.0	14	0.2	21	0.2	0	0.0	0	0.0
Eastern Europe	5361	10.2	5522	9.3	156	3.7	134	2.8	826	10.4	934	9.5	954	26.6	1131	27.4
Albania	11	0.0	5	0.0	0	0.0	0	0.0	11	0.1	9	0.1	1	0.0	1	0.0
Bulgaria	129	0.2	108	0.2	15	0.3	16	0.3	86	1.1	110	1.1	27	0.8	15	0.4
Czechoslovakia	144	0.3	162	0.3	2	0.0	1	0.0	21	0.3	27	0.3	3	0.1	2	0.1
German Dem.Rep.	67	0.1	117	0.2	1	0.0	3	0.1	33	0.4	16	0.2	280	7.8	376	9.1
Hungary	60	0.1	60	0.1	9	0.2	5	0.1	24	0.3	22	0.2	20	0.6	15	0.4
Poland	75	0.1	106	0.2	14	0.3	17	0.3	58	0.7	77	0.8	6	0.2	9	0.2
Romania	41	0.1	45	0.1	2	0.0	1	0.0	29	0.4	30	0.3	3	0.1	3	0.1
7 of E.Europe	527	1.0	603	1.0	43	1.0	43	0.9	262	3.3	291	3.0	341	9.5	421	10.2
USSR	4834	9.2	4919	8.3	113	2.7	91	1.9	564	7.1	643	6.5	614	17.1	710	17.2
E.Europe unallo	-	-	-	-	-	-	-	-	-	-	-	-	-	-	-	-
Unallocated	14	0.0	11	0.0	-	-	-	-	2	0.0	0	0.0	-	-	-	-
World	52684	100.0	59280	100.0	4266	100.0	4801	100.0	7972	100.0	9818	100.0	3589	100.0	4123	100.0

Table II.18. Changes in the geographic orientation of exports of steel products

	Oth West European countries				Eastern European countries				North American countries				Japan			
Years	1983		1984		1983		1984		1983		1984		1983		1984	
Importers	1000t	%	1000t	%	1000t	%	1000t	%	1000t	%	1000t	%	1000t	%	1000t	%
Africa	1455	9.2	1345	7.2	67	0.5	85	0.6	53	1.4	40	1.0	616	2.0	562	1.8
Algeria	591	3.7	411	2.2	29	0.2	18	0.1	24	0.6	1	0.0	4	0.0	23	0.1
Liby.Arab Jam.	271	1.7	39	0.2	2	0.0	2	0.0	2	0.1	1	0.0	48	0.2	22	0.1
Morocco	310	2.0	327	1.7	3	0.0	3	0.0	0	0.0	3	0.1	14	0.0	9	0.0
Tunisia	90	0.6	115	0.6	0	0.0	1	0.0	1	0.0	0	0.0	4	0.0	2	0.0
North Africa	1261	8.0	893	4.8	34	0.2	24	0.2	27	0.7	5	0.1	70	0.2	56	0.2
South Africa	8	0.0	13	0.1	-	-	-	-	3	0.1	2	0.1	87	0.3	100	0.3
Other Africa	185	1.2	439	2.3	34	0.2	61	0.4	23	0.6	33	0.9	458	1.5	406	1.3
Far East	681	4.3	1340	7.1	691	5.0	895	6.0	357	9.4	270	7.1	16539	53.7	17058	53.8
China	163	1.0	593	3.2	247	1.8	352	2.4	30	0.8	5	0.1	6919	22.5	8390	26.5
India	77	0.5	81	0.4	43	0.3	49	0.3	24	0.6	32	0.8	658	2.1	451	1.4
Korea,Rep.of	15	0.1	18	0.1	-	-	-	-	40	1.1	23	0.6	1866	6.1	2225	7.0
3 of Far East	256	1.6	692	3.7	290	2.1	401	2.7	95	2.5	60	1.6	9446	30.6	11066	34.9
Japan	85	0.5	304	1.6	109	0.8	149	1.0	8	0.2	8	0.2	-	-	-	-
Other Far East	340	2.1	344	1.8	293	2.1	344	2.3	255	6.7	202	5.3	7093	23.0	5992	18.9
Middle East	2830	17.9	2697	14.4	1290	9.4	1641	11.1	201	5.3	80	2.1	4915	15.9	3093	9.8
Egypt	267	1.7	561	3.0	342	2.5	334	2.2	43	1.1	35	0.9	108	0.3	126	0.4
Saudi Arabia	217	1.4	293	1.6	28	0.2	72	0.5	86	2.3	32	0.8	2258	7.3	1158	3.7
2 of M.East	484	3.1	854	4.6	370	2.7	406	2.7	130	3.4	66	1.7	2366	7.7	1286	4.1
Other M.East	2346	14.8	1843	9.8	920	6.7	1236	8.3	72	1.9	13	0.3	2549	8.3	1807	5.7
Oceania	9	0.1	27	0.1	-	-	0	0.0	11	0.3	7	0.2	694	2.3	895	2.8
Australia	8	0.1	24	0.1	-	-	0	0.0	5	0.1	6	0.2	353	1.1	447	1.4
New Zealand	1	0.0	2	0.0	-	-	0	0.0	5	0.1	1	0.0	325	1.1	431	1.4
Other Oceania	0	0.0	1	0.0	-	-	0	0.0	1	0.0	0	0.0	16	0.1	18	0.1
North America	1206	7.6	2755	14.7	66	0.5	662	4.5	2701	70.7	2878	75.2	4506	14.6	6297	19.9
Canada	52	0.3	199	1.1	6	0.0	28	0.2	358	9.4	322	8.4	196	0.6	244	0.8
U.S.of America	1154	7.3	2555	13.6	59	0.4	634	4.3	2341	61.3	2555	66.8	4311	14.0	6053	19.1
N.America una.	0	0.0	1	0.0	-	-	-	-	1	0.0	1	0.0	0	0.0	0	0.0
Other America	141	0.9	229	1.2	132	1.0	126	0.9	352	9.2	345	9.0	884	2.9	1097	3.5
Argentina	21	0.1	12	0.1	1	0.0	0	0.0	11	0.3	4	0.1	153	0.5	118	0.4
Brazil	9	0.1	4	0.0	0	0.0	-	-	5	0.1	2	0.1	27	0.1	43	0.1
Mexico	17	0.1	37	0.2	-	-	-	-	169	4.4	226	5.9	82	0.3	243	0.8
Venezuela	5	0.0	8	0.0	-	-	-	-	15	0.4	8	0.2	85	0.3	68	0.2
4 of O.America	52	0.3	61	0.3	1	0.0	0	0.0	200	5.2	240	6.3	347	1.1	471	1.5
Oth.Oth.America	89	0.6	168	0.9	131	1.0	126	0.8	153	4.0	104	2.7	537	1.7	626	2.0
Western Europe	7566	47.8	8150	43.5	3762	27.4	4281	28.8	140	3.7	205	5.4	607	2.0	582	1.8
B.L.E.U.	277	1.7	257	1.4	156	1.1	142	1.0	8	0.2	4	0.1	54	0.2	58	0.2
Denmark	483	3.1	528	2.8	106	0.8	109	0.7	0	0.0	1	0.0	8	0.0	6	0.0
France	567	3.6	527	2.8	244	1.8	338	2.3	11	0.3	20	0.5	19	0.1	9	0.0
Germany,Fed.Rep	2579	16.3	2671	14.3	876	6.4	1034	7.0	10	0.3	30	0.8	93	0.3	47	0.1
Ireland	33	0.2	52	0.3	3	0.0	8	0.1	1	0.0	1	0.0	1	0.0	0	0.0
Italy	679	4.3	862	4.6	449	3.3	420	2.8	66	1.7	92	2.4	26	0.1	22	0.1
Netherlands	438	2.8	427	2.3	63	0.5	67	0.5	3	0.1	3	0.1	21	0.1	13	0.0
United Kingdom	648	4.1	693	3.7	169	1.2	156	1.0	25	0.7	20	0.5	56	0.2	88	0.3
EEC(9)	5704	36.0	6017	32.1	2065	15.1	2273	15.3	124	3.3	171	4.5	277	0.9	244	0.8
Finland	222	1.4	197	1.1	131	1.0	93	0.6	0	0.0	1	0.0	6	0.0	4	0.0
Iceland	26	0.2	22	0.1	1	0.0	2	0.0	1	0.0	0	0.0	-	-	0	0.0
Norway	323	2.0	334	1.8	43	0.3	53	0.4	0	0.0	1	0.0	65	0.2	64	0.2
Sweden	418	2.6	390	2.1	159	1.2	125	0.8	2	0.1	4	0.1	30	0.1	20	0.1
Northern Europe	989	6.2	943	5.0	334	2.4	274	1.8	4	0.1	7	0.2	100	0.3	87	0.3
Greece	65	0.4	77	0.4	75	0.5	84	0.6	0	0.0	17	0.4	81	0.3	131	0.4
Portugal	62	0.4	86	0.5	43	0.3	31	0.2	0	0.0	0	0.0	8	0.0	0	0.0
Spain	52	0.3	31	0.2	58	0.4	37	0.3	7	0.2	3	0.1	13	0.0	11	0.0
Turkey	189	1.2	466	2.5	148	1.1	324	2.2	2	0.0	7	0.2	100	0.3	75	0.2
Yugoslavia	143	0.9	133	0.7	732	5.3	810	5.5	2	0.0	0	0.0	7	0.0	5	0.0
Southern Europe	510	3.2	793	4.2	1056	7.7	1286	8.7	11	0.3	26	0.7	208	0.7	232	0.7
Austria	121	0.8	128	0.7	170	1.2	222	1.5	0	0.0	0	0.0	4	0.0	6	0.0
Switzerland	211	1.3	201	1.1	86	0.6	137	0.9	1	0.0	2	0.0	17	0.1	13	0.0
Central Europe	332	2.1	329	1.8	256	1.9	359	2.4	1	0.0	2	0.0	21	0.1	19	0.1
W.Europe unallo	32	0.2	68	0.4	51	0.4	89	0.6	0	0.0	0	0.0	1	0.0	0	0.0
Eastern Europe	1937	12.2	2199	11.7	3376	24.6	3420	23.0	2	0.1	1	0.0	2060	6.7	2098	6.6
Albania	12	0.1	9	0.0	75	0.5	60	0.4	-	-	-	-	5	0.0	3	0.0
Bulgaria	128	0.8	140	0.7	229	1.7	296	2.0	-	-	0	0.0	11	0.0	3	0.0
Czechoslovakia	25	0.2	31	0.2	136	1.0	218	1.5	-	-	-	-	1	0.0	1	0.0
German Dem.Rep.	315	2.0	395	2.1	465	3.4	434	2.9	-	-	-	-	8	0.0	2	0.0
Hungary	53	0.3	42	0.2	94	0.7	105	0.7	0	0.0	0	0.0	2	0.0	1	0.0
Poland	78	0.5	103	0.6	433	3.2	512	3.4	0	0.0	0	0.0	9	0.0	8	0.0
Romania	34	0.2	34	0.2	181	1.3	176	1.2	1	0.0	-	-	12	0.0	9	0.0
7 of E.Europe	645	4.1	755	4.0	1614	11.8	1801	12.1	1	0.0	0	0.0	48	0.2	26	0.1
USSR	1291	8.2	1444	7.7	1762	12.9	1619	10.9	1	0.0	1	0.0	2012	6.5	2072	6.5
E.Europe unallo	-	-	-	-	-	-	-	-	-	-	-	-	-	-	-	-
Unallocated	2	0.0	0	0.0	4323	31.5	3735	25.2	-	-	-	-	-	-	-	-
World	15827	100.0	18741	100.0	13707	100.0	14846	100.0	3819	100.0	3826	100.0	30821	100.0	31684	100.0

Table II.19. Changes in the geographic orientation of exports of steel products

	Oceania			
Years	1983		1984	
Importers	1000t	%	1000t	%
Africa	2	0.2	1	0.0
Algeria	-	-	-	-
Liby.Arab Jam.	-	-	-	-
Morocco	-	-	-	-
Tunisia	-	-	-	-
North Africa	-	-	-	-
South Africa	0	0.0	0	0.0
Other Africa	2	0.2	0	0.0
Far East	539	46.2	570	49.7
China	156	13.4	224	19.6
India	16	1.4	8	0.7
Korea,Rep.of	20	1.8	36	3.2
3 of Far East	193	16.5	269	23.5
Japan	50	4.2	34	3.0
Other Far East	297	25.4	267	23.3
Middle East	94	8.0	108	9.5
Egypt	6	0.5	-	-
Saudi Arabia	37	3.2	1	0.1
2 of M.East	43	3.7	1	0.1
Other M.East	51	4.3	107	9.4
Oceania	184	15.8	210	18.3
Australia	18	1.6	16	1.4
New Zealand	102	8.7	118	10.3
Other Oceania	64	5.5	76	6.6
North America	177	15.2	204	17.8
Canada	13	1.1	9	0.8
U.S.of America	164	14.0	195	17.1
N.America una.	0	0.0	-	-
Other America	28	2.4	22	1.9
Argentina	21	1.8	21	1.9
Brazil	-	-	-	-
Mexico	-	-	0	0.0
Venezuela	0	0.0	-	-
4 of O.America	21	1.8	21	1.9
Oth.Oth.America	7	0.6	1	0.1
Western Europe	115	9.8	1	0.1
B.L.E.U.	9	0.8	-	-
Denmark	-	-	-	-
France	0	0.0	-	-
Germany,Fed.Rep	10	0.9	0	0.0
Ireland	-	-	-	-
Italy	22	1.9	-	-
Netherlands	7	0.6	0	0.0
United Kingdom	3	0.2	1	0.1
EEC(9)	51	4.4	1	0.1
Finland	-	-	-	-
Iceland	-	-	0	0.0
Norway	-	-	-	-
Sweden	-	-	-	-
Northern Europe	-	-	0	0.0
Greece	64	5.5	0	0.0
Portugal	-	-	0	0.0
Spain	-	-	-	-
Turkey	-	-	-	-
Yugoslavia	-	-	-	-
Southern Europe	64	5.5	0	0.0
Austria	-	-	-	-
Switzerland	0	0.0	0	0.0
Central Europe	0	0.0	0	0.0
W.Europe unallo	-	-	-	-
Eastern Europe	-	-	-	-
Albania	-	-	-	-
Bulgaria	-	-	-	-
Czechoslovakia	-	-	-	-
German Dem.Rep.	-	-	-	-
Hungary	-	-	-	-
Poland	-	-	-	-
Romania	-	-	-	-
7 of E.Europe	-	-	-	-
USSR	-	-	-	-
E.Europe unallo	-	-	-	-
Unallocated	28	2.4	30	2.6
World	1168	100.0	1146	100.0

Table II.20. Changes in the geographic orientation of exports of steel products excluding intra-regional trade

Countries	European Economic Community				North European countries				South European countries				Central European countries			
Years	1983		1984		1983		1984		1983		1984		1983		1984	
	1000t	%	1000t	%	1000t	%	1000t	%	1000t	%	1000t	%	1000t	%	1000t	%
Africa	2063	7.6	2267	7.2	42	1.2	46	1.1	1367	17.7	1227	13.3	45	1.3	71	1.8
Algeria	485	1.8	603	1.9	19	0.5	20	0.5	532	6.9	327	3.5	39	1.1	64	1.6
Liby.Arab Jam.	236	0.9	246	0.8	3	0.1	2	0.0	266	3.4	34	0.4	2	0.0	4	0.1
Morocco	215	0.8	340	1.1	0	0.0	0	0.0	310	4.0	327	3.5	0	0.0	0	0.0
Tunisia	198	0.7	194	0.6	0	0.0	0	0.0	90	1.2	115	1.2	1	0.0	0	0.0
North Africa	1135	4.2	1383	4.4	23	0.6	22	0.5	1197	15.5	803	8.7	41	1.2	68	1.7
South Africa	73	0.3	90	0.3	4	0.1	8	0.2	2	0.0	3	0.0	2	0.1	1	0.0
Other Africa	855	3.1	794	2.5	15	0.4	16	0.4	169	2.2	421	4.6	2	0.1	2	0.1
Far East	3284	12.1	3582	11.4	103	2.9	250	6.2	566	7.3	1069	11.6	12	0.4	20	0.5
China	1220	4.5	1256	4.0	5	0.1	133	3.3	157	2.0	447	4.8	1	0.0	13	0.3
India	502	1.8	929	3.0	20	0.6	20	0.5	54	0.7	58	0.6	4	0.1	3	0.1
Korea,Rep.of	60	0.2	133	0.4	4	0.1	7	0.2	12	0.2	10	0.1	0	0.0	0	0.0
3 of Far East	1782	6.6	2318	7.4	28	0.8	160	3.9	223	2.9	515	5.6	5	0.2	16	0.4
Japan	7	0.0	28	0.1	5	0.2	8	0.2	78	1.0	296	3.2	2	0.0	0	0.0
Other Far East	1495	5.5	1236	3.9	70	2.0	83	2.0	265	3.4	258	2.8	5	0.2	3	0.1
Middle East	2840	10.5	2805	9.0	129	3.6	49	1.2	2571	33.3	2544	27.5	131	3.8	104	2.6
Egypt	338	1.2	292	0.9	7	0.2	8	0.2	250	3.2	535	5.8	10	0.3	18	0.5
Saudi Arabia	512	1.9	554	1.8	51	1.4	22	0.5	164	2.1	267	2.9	2	0.0	4	0.1
2 of M.East	850	3.1	846	2.7	58	1.6	30	0.7	414	5.4	802	8.7	12	0.3	22	0.6
Other M.East	1990	7.3	1959	6.3	71	2.0	19	0.5	2157	28.0	1742	18.9	119	3.5	82	2.1
Oceania	126	0.5	116	0.4	6	0.2	11	0.3	1	0.0	13	0.1	2	0.1	3	0.1
Australia	72	0.3	48	0.2	5	0.1	9	0.2	1	0.0	12	0.1	2	0.1	3	0.1
New Zealand	33	0.1	38	0.1	1	0.0	1	0.0	0	0.0	0	0.0	0	0.0	0	0.0
Other Oceania	22	0.1	30	0.1	0	0.0	0	0.0	0	0.0	1	0.0	0	0.0	0	0.0
North America	4479	16.5	6782	21.7	472	13.3	971	23.9	706	9.2	1633	17.7	28	0.8	151	3.8
Canada	373	1.4	667	2.1	21	0.6	68	1.7	26	0.3	114	1.2	5	0.2	17	0.4
U.S.of America	4102	15.1	6019	19.2	450	12.7	903	22.2	681	8.8	1518	16.4	23	0.7	135	3.4
N.America una.	5	0.0	96	0.3	-	-	-	-	0	0.0	1	0.0	-	-	-	-
Other America	1116	4.1	1220	3.9	17	0.5	39	1.0	113	1.5	176	1.9	11	0.3	14	0.3
Argentina	328	1.2	220	0.7	6	0.2	8	0.2	14	0.2	4	0.0	1	0.0	1	0.0
Brazil	48	0.2	57	0.2	3	0.1	3	0.1	1	0.0	-	-	6	0.2	1	0.0
Mexico	184	0.7	367	1.2	5	0.1	6	0.1	11	0.1	29	0.3	1	0.0	2	0.0
Venezuela	92	0.3	74	0.2	1	0.0	1	0.0	4	0.1	7	0.1	0	0.0	1	0.0
4 of O.America	652	2.4	718	2.3	14	0.4	18	0.4	31	0.4	39	0.4	7	0.2	4	0.1
Oth.Oth.America	465	1.7	502	1.6	4	0.1	21	0.5	82	1.1	137	1.5	3	0.1	10	0.2
Western Europe	7887	29.0	8994	28.7	2624	73.9	2571	63.2	1559	20.2	1644	17.8	2244	65.5	2460	62.2
B.L.E.U.	.		.		121	3.4	103	2.5	124	1.6	115	1.2	32	0.9	39	1.0
Denmark	.		.		426	12.0	473	11.6	9	0.1	6	0.1	48	1.4	50	1.3
France	.		.		181	5.1	156	3.8	250	3.2	227	2.5	135	3.9	144	3.6
Germany,Fed.Rep	.		.		861	24.3	804	19.8	412	5.3	518	5.6	1306	38.1	1349	34.1
Ireland	.		.		15	0.4	31	0.8	13	0.2	18	0.2	5	0.1	3	0.1
Italy	.		.		96	2.7	126	3.1	211	2.7	260	2.8	372	10.9	475	12.0
Netherlands	.		.		294	8.3	266	6.5	62	0.8	59	0.6	81	2.4	101	2.6
United Kingdom	.		..		403	11.4	412	10.1	172	2.2	199	2.2	73	2.1	83	2.1
EEC(9)	-		-		2396	67.5	2371	58.3	1254	16.3	1401	15.2	2053	59.9	2244	56.7
Finland	249	0.9	280	0.9	.		.		51	0.7	38	0.4	12	0.4	13	0.3
Iceland	16	0.1	19	0.1	.		.		11	0.1	4	0.0	0	0.0	0	0.0
Norway	947	3.5	848	2.7	.		.		35	0.5	31	0.3	12	0.4	15	0.4
Sweden	1301	4.8	1414	4.5	.		.		85	1.1	43	0.5	64	1.9	68	1.7
Northern Europe	2513	9.3	2561	8.2	.		.		183	2.4	117	1.3	89	2.6	96	2.4
Greece	1015	3.7	1164	3.7	45	1.3	44	1.1	.		.		3	0.1	5	0.1
Portugal	357	1.3	414	1.3	11	0.3	14	0.4	.		.		3	0.1	2	0.1
Spain	1003	3.7	1218	3.9	40	1.1	19	0.5	.		.		9	0.3	11	0.3
Turkey	359	1.3	722	2.3	38	1.1	6	0.2	.		.		7	0.2	12	0.3
Yugoslavia	280	1.0	363	1.2	13	0.4	14	0.3	.		.		80	2.3	88	2.2
Southern Europe	3014	11.1	3880	12.4	147	4.1	97	2.4	.		.		102	3.0	119	3.0
Austria	675	2.5	749	2.4	24	0.7	22	0.5	42	0.5	56	0.6	.		.	
Switzerland	1624	6.0	1733	5.5	39	1.1	33	0.8	66	0.9	50	0.5	.		.	
Central Europe	2299	8.5	2481	7.9	63	1.8	55	1.4	108	1.4	106	1.1	.		.	
W.Europe unallo	61	0.2	72	0.2	18	0.5	47	1.2	14	0.2	21	0.2	0	0.0	0	0.0
Eastern Europe	5361	19.7	5522	17.6	156	4.4	134	3.3	826	10.7	934	10.1	954	27.8	1131	28.6
Albania	11	0.0	5	0.0	0	0.0	0	0.0	11	0.1	9	0.1	1	0.0	1	0.0
Bulgaria	129	0.5	108	0.3	15	0.4	16	0.4	86	1.1	110	1.2	27	0.8	15	0.4
Czechoslovakia	144	0.5	162	0.5	2	0.0	1	0.0	21	0.3	27	0.3	3	0.1	2	0.1
German Dem.Rep.	67	0.2	117	0.4	1	0.0	3	0.1	33	0.4	16	0.2	280	8.2	376	9.5
Hungary	60	0.2	60	0.2	9	0.3	5	0.1	24	0.3	22	0.2	20	0.6	15	0.4
Poland	75	0.3	106	0.3	14	0.4	17	0.4	58	0.8	77	0.8	6	0.2	9	0.2
Romania	41	0.2	45	0.1	2	0.1	1	0.0	29	0.4	30	0.3	3	0.1	1	0.0
7 of E.Europe	527	1.9	603	1.9	43	1.2	43	1.0	262	3.4	291	3.1	341	9.9	421	10.7
USSR	4834	17.8	4919	15.7	113	3.2	91	2.2	564	7.3	643	7.0	614	17.9	710	18.0
E.Europe unallo	-		-		-	-	-	-	-		-		-		-	
Unallocated	14	0.1	11	0.0	-	-	-	-	2	0.0	0	0.0	-		-	
World	27170	100.0	31300	100.0	3549	100.0	4071	100.0	7711	100.0	9241	100.0	3427	100.0	3955	100.0

Table II.21. Changes in the geographic orientation of exports of steel products

Countries	North American countries 1983 1000t	%	1984 1000t	%	Japan 1983 1000t	%	1984 1000t	%	East European countries 1983 1000t	%	1984 1000t	%	Oceania 1983 1000t	%	1984 1000t	%
Africa	53	4.7	40	4.2	616	2.0	562	1.8	67	0.6	85	0.6	2	0.2	1	0.1
Algeria	24	2.2	1	0.1	4	0.0	23	0.1	29	0.2	18	0.1	-	-	-	-
Liby.Arab Jam.	2	0.2	1	0.1	48	0.2	22	0.1	2	0.0	2	0.0	-	-	-	-
Morocco	0	0.0	3	0.3	14	0.0	9	0.0	3	0.0	3	0.0	-	-	-	-
Tunisia	1	0.1	0	0.0	4	0.0	2	0.0	0	0.0	1	0.0	-	-	-	-
North Africa	27	2.4	5	0.5	70	0.2	56	0.2	34	0.3	24	0.2	-	-	-	-
South Africa	3	0.2	2	0.2	87	0.3	100	0.3	-	-	-	-	0	0.0	0	0.0
Other Africa	23	2.1	33	3.5	458	1.5	406	1.3	34	0.3	61	0.5	2	0.2	0	0.0
Far East	357	32.0	270	28.5	16539	53.7	17058	53.8	691	5.7	895	6.9	539	54.8	570	60.9
China	30	2.7	5	0.5	6919	22.5	8390	26.5	247	2.0	352	2.7	156	15.9	224	24.0
India	24	2.2	32	3.4	658	2.1	451	1.4	43	0.4	49	0.4	16	1.7	8	0.9
Korea,Rep.of	40	3.6	23	2.5	1868	6.1	2225	7.0	-	-	-	-	20	2.1	36	3.9
3 of Far East	95	8.5	60	6.4	9446	30.6	11066	34.9	290	2.4	401	3.1	193	19.6	269	28.7
Japan	8	0.7	8	0.8	-	-	-	-	109	0.9	149	1.1	50	5.0	34	3.6
Other Far East	255	22.8	202	21.3	7093	23.0	5992	18.9	293	2.4	344	2.6	297	30.2	267	28.5
Middle East	201	18.0	80	8.4	4915	15.9	3093	9.8	1290	10.7	1641	12.6	94	9.5	108	11.6
Egypt	43	3.9	35	3.6	108	0.3	128	0.4	342	2.8	334	2.6	6	0.6	-	-
Saudi Arabia	86	7.7	32	3.4	2258	7.3	1158	3.7	28	0.2	72	0.6	37	3.8	1	0.1
2 of M.East	130	11.6	66	7.0	2366	7.7	1286	4.1	370	3.1	406	3.1	43	4.4	1	0.1
Other M.East	72	6.4	13	1.4	2549	8.3	1807	5.7	920	7.6	1236	9.5	51	5.1	107	11.5
Oceania	11	1.0	7	0.8	694	2.3	895	2.8	-	-	0	0.0	.	.	.	.
Australia	5	0.5	6	0.6	353	1.1	447	1.4	-	-	0	0.0	.	.	.	.
New Zealand	5	0.4	1	0.1	325	1.1	431	1.4	-	-	0	0.0	.	.	.	.
Other Oceania	1	0.1	0	0.0	16	0.1	18	0.1	-	-	0	0.0	.	.	.	.
North America	.	.	.	.	4506	14.6	6297	19.9	66	0.5	662	5.1	177	18.0	204	21.8
Canada	.	.	.	.	196	0.6	244	0.8	6	0.1	28	0.2	13	1.4	9	1.0
U.S.of America	.	.	.	.	4311	14.0	6053	19.1	59	0.5	634	4.9	164	16.7	195	20.9
N.America una.	.	.	.	.	0	0.0	0	0.0	-	-	-	-	0	0.0	-	-
Other America	352	31.5	345	36.4	884	2.9	1097	3.5	132	1.1	126	1.0	28	2.8	22	2.3
Argentina	11	1.0	4	0.5	153	0.5	118	0.4	1	0.0	0	0.0	21	2.1	21	2.3
Brazil	5	0.4	2	0.3	27	0.1	43	0.1	0	0.0	-	-	-	-	-	-
Mexico	169	15.1	226	23.8	82	0.3	243	0.8	-	-	-	-	-	-	0	0.0
Venezuela	15	1.3	8	0.8	85	0.3	68	0.2	-	-	-	-	0	0.0	-	-
4 of O.America	200	17.9	240	25.3	347	1.1	471	1.5	1	0.0	0	0.0	21	2.1	21	2.3
Oth.Oth.America	153	13.6	104	11.0	537	1.7	626	2.0	131	1.1	126	1.0	7	0.7	1	0.1
Western Europe	140	12.6	205	21.7	607	2.0	582	1.8	3762	31.1	4281	32.8	115	11.7	1	0.1
B.L.E.U.	8	0.7	4	0.5	54	0.2	58	0.2	156	1.3	142	1.1	9	0.9	-	-
Denmark	0	0.0	1	0.1	8	0.0	6	0.0	106	0.9	109	0.8	-	-	-	-
France	11	1.0	20	2.1	19	0.1	9	0.0	244	2.0	338	2.6	0	0.0	-	-
Germany,Fed.Rep	10	0.9	30	3.2	93	0.3	47	0.1	876	7.2	1034	7.9	10	1.0	0	0.0
Ireland	1	0.1	1	0.1	1	0.0	0	0.0	3	0.0	8	0.1	-	-	-	-
Italy	66	5.9	92	9.7	26	0.1	22	0.1	449	3.7	420	3.2	22	2.2	0	0.0
Netherlands	3	0.3	3	0.3	21	0.1	13	0.0	63	0.5	67	0.5	7	0.7	0	0.0
United Kingdom	25	2.2	20	2.1	56	0.2	88	0.3	169	1.4	156	1.2	3	0.3	1	0.1
EEC(9)	124	11.1	171	18.0	277	0.9	244	0.8	2065	17.1	2273	17.4	51	5.2	1	0.1
Finland	0	0.0	1	0.1	6	0.0	4	0.0	131	1.1	93	0.7	-	-	-	-
Iceland	1	0.1	0	0.0	-	-	0	0.0	1	0.0	2	0.0	-	-	0	0.0
Norway	0	0.0	1	0.1	65	0.2	64	0.2	43	0.4	53	0.4	-	-	-	-
Sweden	2	0.2	4	0.4	30	0.1	20	0.1	159	1.3	125	1.0	-	-	-	-
Northern Europe	4	0.4	7	0.7	100	0.3	87	0.3	334	2.8	274	2.1	-	-	0	0.0
Greece	0	0.0	17	1.8	81	0.3	131	0.4	75	0.6	84	0.6	64	6.5	0	0.0
Portugal	0	0.0	0	0.0	8	0.0	9	0.0	43	0.4	31	0.2	-	-	0	0.0
Spain	7	0.6	3	0.3	13	0.0	11	0.0	58	0.5	37	0.3	-	-	-	-
Turkey	2	0.1	7	0.7	100	0.3	75	0.2	148	1.2	324	2.5	-	-	-	-
Yugoslavia	2	0.2	0	0.0	7	0.0	5	0.0	732	6.1	810	6.2	-	-	-	-
Southern Europe	11	0.9	26	2.7	208	0.7	232	0.7	1056	8.7	1286	9.9	64	6.5	0	0.0
Austria	0	0.0	0	0.0	4	0.0	6	0.0	170	1.4	222	1.7	-	-	-	-
Switzerland	1	0.1	2	0.2	17	0.1	13	0.0	86	0.7	137	1.0	0	0.0	0	0.0
Central Europe	1	0.1	2	0.2	21	0.1	19	0.1	256	2.1	359	2.8	0	0.0	0	0.0
W.Europe unallo	0	0.0	0	0.0	1	0.0	0	0.0	51	0.4	89	0.7	-	-	-	-
Eastern Europe	2	0.2	1	0.1	2060	6.7	2098	6.6	1762	14.6	1619	12.4	-	-	-	-
Albania	-	-	-	-	5	0.0	3	0.0	.	.	.	.	-	-	-	-
Bulgaria	-	-	0	0.0	11	0.0	3	0.0	.	.	.	.	-	-	-	-
Czechoslovakia	-	-	-	-	1	0.0	1	0.0	.	.	.	.	-	-	-	-
German Dem.Rep.	-	-	-	-	8	0.0	2	0.0	.	.	.	.	-	-	-	-
Hungary	0	0.0	0	0.0	2	0.0	1	0.0	.	.	.	.	-	-	-	-
Poland	0	0.0	0	0.0	9	0.0	8	0.0	.	.	.	.	-	-	-	-
Romania	1	0.1	-	-	12	0.0	9	0.0	.	.	.	.	-	-	-	-
7 of E.Europe	1	0.1	0	0.0	48	0.2	26	0.1	.	.	.	.	-	-	-	-
USSR	1	0.1	1	0.1	2012	6.5	2072	6.5	1762	14.6	1619	12.4	.	.	.	.
E.Europe unallo	-	-	-	-	-	-	-	-	.	.	.	.	-	-	-	-
Unallocated	-	-	-	-	-	-	-	-	4323	35.7	3735	28.6	28	2.9	30	3.2
World	1118	100.0	948	100.0	30821	100.0	31684	100.0	12093	100.0	13045	100.0	984	100.0	936	100.0

Table II.22. Changes in the geographic orientation of exports of steel products

Years / Countries	Oth West European countries 1983 1000t	1983 %	1984 1000t	1984 %
Africa	1455	9.2	1345	7.2
Algeria	591	3.7	411	2.2
Liby.Arab Jam.	271	1.7	39	0.2
Morocco	310	2.0	327	1.7
Tunisia	90	0.6	115	0.6
North Africa	1261	8.0	893	4.8
South Africa	8	0.0	13	0.1
Other Africa	185	1.2	439	2.3
Far East	681	4.3	1340	7.1
China	163	1.0	593	3.2
India	77	0.5	81	0.4
Korea,Rep.of	15	0.1	18	0.1
3 of Far East	256	1.6	692	3.7
Japan	85	0.5	304	1.6
Other Far East	340	2.1	344	1.8
Middle East	2830	17.9	2697	14.4
Egypt	267	1.7	561	3.0
Saudi Arabia	217	1.4	293	1.6
2 of M.East	484	3.1	854	4.6
Other M.East	2346	14.8	1843	9.8
Oceania	9	0.1	27	0.1
Australia	8	0.1	24	0.1
New Zealand	1	0.0	2	0.0
Other Oceania	0	0.0	1	0.0
North America	1206	7.6	2755	14.7
Canada	52	0.3	199	1.1
U.S.of America	1154	7.3	2555	13.6
N.America una.	0	0.0	1	0.0
Other America	141	0.9	229	1.2
Argentina	21	0.1	12	0.1
Brazil	9	0.1	4	0.0
Mexico	17	0.1	37	0.2
Venezuela	5	0.0	8	0.0
4 of O.America	52	0.3	61	0.3
Oth.Oth.America	89	0.6	168	0.9
Western Europe	7566	47.8	8150	43.5
B.L.E.U.	277	1.7	257	1.4
Denmark	483	3.1	528	2.8
France	567	3.6	527	2.8
Germany,Fed.Rep	2579	16.3	2671	14.3
Ireland	33	0.2	52	0.3
Italy	679	4.3	862	4.6
Netherlands	438	2.8	427	2.3
United Kingdom	648	4.1	693	3.7
EEC(9)	5704	36.0	6017	32.1
Finland	222	1.4	197	1.1
Iceland	26	0.2	22	0.1
Norway	323	2.0	334	1.8
Sweden	418	2.6	390	2.1
Northern Europe	989	6.2	943	5.0
Greece	65	0.4	77	0.4
Portugal	62	0.4	86	0.5
Spain	52	0.3	31	0.2
Turkey	189	1.2	466	2.5
Yugoslavia	143	0.9	133	0.7
Southern Europe	510	3.2	793	4.2
Austria	121	0.8	128	0.7
Switzerland	211	1.3	201	1.1
Central Europe	332	2.1	329	1.8
W.Europe unallo	32	0.2	68	0.4
Eastern Europe	1937	12.2	2199	11.7
Albania	12	0.1	9	0.0
Bulgaria	128	0.8	140	0.7
Czechoslovakia	25	0.2	31	0.2
German Dem.Rep.	315	2.0	395	2.1
Hungary	53	0.3	42	0.2
Poland	78	0.5	103	0.6
Romania	34	0.2	34	0.2
7 of E.Europe	645	4.1	755	4.0
USSR	1291	8.2	1444	7.7
E.Europe unallo	-	-	-	-
Unallocated	2	0.0	0	0.0
World	15827	100.0	18741	100.0

Table II.23. Structural changes of exports of steel products
(Percentage of region exports)

Years	European Economic Community				North European countries				South European countries				Central European countries			
	1983		1984		1983		1984		1983		1984		1983		1984	
Products	1000t	%	1000t	%	1000t	%	1000t	%	1000t	%	1000t	%	1000t	%	1000t	%
Ingots & semis	4956	18.2	6680	21.3	648	18.3	937	23.0	908	11.8	1045	11.3	354	10.3	480	12.1
Long Products	5757	21.2	6331	20.2	991	27.9	1141	28.0	4739	61.5	5438	58.9	859	25.1	1045	26.4
Flat Products	9823	36.2	11035	35.3	1596	45.0	1630	40.0	1222	15.9	1581	17.1	1639	47.8	1700	43.0
Wire	544	2.0	626	2.0	63	1.8	71	1.7	69	0.9	102	1.1	70	2.0	81	2.1
Tubes & Fitting	6091	22.4	6629	21.2	251	7.1	292	7.2	773	10.0	1074	11.6	505	14.7	649	16.4
Total	27170	100.0	31300	100.0	3549	100.0	4071	100.0	7711	100.0	9241	100.0	3427	100.0	3955	100.0

Table II.24. Structural changes of exports of steel products
(Percentage of region exports)

Years	Oth West European countries				Eastern European countries				North America				Japan			
	1983		1984		1983		1984		1983		1984		1983		1984	
Products	1000t	%	1000t	%	1000t	%	1000t	%	1000t	%	1000t	%	1000t	%	1000t	%
Ingots & semis	2067	13.1	2758	14.7	1115	9.2	1088	8.3	98	8.8	59	6.2	3301	10.7	3331	10.5
Long Products	6976	44.1	8152	43.5	5160	42.7	5451	41.8	244	21.9	205	21.6	8432	27.4	7873	24.8
Flat Products	4890	30.9	5367	28.6	4309	35.6	4861	37.3	555	49.6	514	54.3	13574	44.0	14044	44.3
Wire	235	1.5	295	1.6	479	4.0	471	3.6	13	1.1	12	1.3	314	1.0	354	1.1
Tubes & Fitting	1658	10.5	2169	11.6	1032	8.5	1174	9.0	208	18.6	158	16.6	5200	16.9	6083	19.2
Total	15827	100.0	18741	100.0	12093	100.0	13045	100.0	1118	100.0	948	100.0	30821	100.0	31684	100.0

Table II.25. Structural changes of exports of steel products
(Percentage of region exports)

Years	Oceania			
	1983		1984	
Products	1000t	%	1000t	%
Ingots & semis	343	34.9	452	48.3
Long Products	163	16.5	85	9.1
Flat Products	442	45.0	359	38.4
Wire	6	0.6	8	0.8
Tubes & Fitting	29	3.0	32	3.4
Total	984	100.0	936	100.0

Table II.26. Change in exports of all products by region or country

Importers	Economic European community 1983 1000t	1983 %	1984 1000t	1984 %	North European countries 1983 1000t	1983 %	1984 1000t	1984 %	South European countries 1983 1000t	1983 %	1984 1000t	1984 %	Central European countries 1983 1000t	1983 %	1984 1000t	1984 %
Africa	2063	7.6	2267	7.2	42	1.2	46	1.1	1367	17.7	1227	13.3	45	1.3	71	1.8
North Africa	1135	4.2	1383	4.4	23	0.6	22	0.5	1197	15.5	803	8.7	41	1.2	68	1.7
South Africa	73	0.3	90	0.3	4	0.1	8	0.2	2	0.0	3	0.0	2	0.1	1	0.0
Other Africa	855	3.1	794	2.5	15	0.4	16	0.4	169	2.2	421	4.6	2	0.1	2	0.1
Far East	3284	12.1	3582	11.4	103	2.9	250	6.2	566	7.3	1069	11.6	12	0.4	20	0.5
3 of Far East	1782	6.6	2318	7.4	28	0.8	160	3.9	223	2.9	515	5.6	5	0.2	16	0.4
Japan	7	0.0	28	0.1	5	0.2	8	0.2	78	1.0	296	3.2	2	0.0	0	0.0
Other Far East	1495	5.5	1236	3.9	70	2.0	83	2.0	265	3.4	258	2.8	5	0.2	3	0.1
Middle East	2840	10.5	2805	9.0	129	3.6	49	1.2	2571	33.3	2544	27.5	131	3.8	104	2.6
2 of Middle Eas	850	3.1	846	2.7	58	1.6	30	0.7	414	5.4	802	8.7	12	0.3	22	0.6
Other M.East	1990	7.3	1959	6.3	71	2.0	19	0.5	2157	28.0	1742	18.9	119	3.5	82	2.1
Oceania	126	0.5	116	0.4	6	0.2	11	0.3	1	0.0	13	0.1	2	0.1	3	0.1
Other Oceania	22	0.1	30	0.1	0	0.0	0	0.0	0	0.0	1	0.0	0	0.0	0	0.0
North America	4479	16.5	6782	21.7	472	13.3	971	23.9	706	9.2	1633	17.7	28	0.8	151	3.8
N.America una.	5	0.0	96	0.3	-	-	-	-	0	0.0	1	0.0	-	-	-	-
Other America	1116	4.1	1220	3.9	17	0.5	39	1.0	113	1.5	176	1.9	11	0.3	14	0.3
4 of O.America	652	2.4	718	2.3	14	0.4	18	0.4	31	0.4	39	0.4	7	0.2	4	0.1
Oth.Oth.America	465	1.7	502	1.6	4	0.1	21	0.5	82	1.1	137	1.5	3	0.1	10	0.2
Western Europe	7887	29.0	8994	28.7	2624	73.9	2571	63.2	1559	20.2	1644	17.8	2244	65.5	2460	62.2
EEC(9)					2396	67.5	2371	58.3	1254	16.3	1401	15.2	2053	59.9	2244	56.7
Northern Europe	2513	9.3	2561	8.2	147	4.1	97	2.4	183	2.4	117	1.3	89	2.6	96	2.4
Southern Europe	3014	11.1	3880	12.4	63	1.8	55	1.4	108	1.4	106	1.1	102	3.0	119	3.0
Central Europe	2299	8.5	2481	7.9	18	0.5	47	1.2	14	0.2	21	0.2	·	0.0	·	·
W.Europe unallo	61	0.2	72	0.2									0	0.0	0	0.0
Eastern Europe	5361	19.7	5522	17.6	156	4.4	134	3.3	826	10.7	934	10.1	954	27.8	1131	28.6
7 of E.Europe	527	1.9	603	1.9	43	1.2	43	1.0	262	3.4	291	3.1	341	9.9	421	10.7
USSR	4834	17.8	4919	15.7	113	3.2	91	2.2	564	7.3	643	7.0	614	17.9	710	18.0
E.Europe unallo	-	-			-	-	-	-	-	-	-	-	-	-	-	-
Unallocated	14	0.1	11	0.0	-	-	-	-	2	0.0	0	0.0	0	0.0	-	-
World	27170	100.0	31300	100.0	3549	100.0	4071	100.0	7711	100.0	9241	100.0	3427	100.0	3955	100.0

Table II.27. Change in exports of all products by region or country

Importers	Oth West European countries 1983 1000t	%	1984 1000t	%	East European countries 1983 1000t	%	1984 1000t	%	North American countries 1983 1000t	%	1984 1000t	%	Japan 1983 1000t	%	1984 1000t	%
Africa	1455	9.2	1345	7.2	67	0.6	85	0.6	53	4.7	40	4.2	616	2.0	562	1.8
North Africa	1261	8.0	893	4.8	34	0.3	24	0.2	27	2.4	5	0.5	70	0.2	56	0.2
South Africa	8	0.0	13	0.1	-	-	-	-	3	0.2	2	0.2	87	0.3	100	0.3
Other Africa	185	1.2	439	2.3	34	0.3	61	0.5	23	2.1	33	3.5	458	1.5	406	1.3
Far East	681	4.3	1340	7.1	691	5.7	895	6.9	357	32.0	270	28.5	16539	53.7	17058	53.8
3 of Far East	256	1.6	692	3.7	290	2.4	401	3.1	95	8.5	60	6.4	9446	30.6	11066	34.9
Japan	85	0.5	304	1.6	109	0.9	149	1.1	8	0.7	8	0.8	-	-	-	-
Other Far East	340	2.1	344	1.8	293	2.4	344	2.6	255	22.8	202	21.3	7093	23.0	5992	18.9
Middle East	2830	17.9	2697	14.4	1290	10.7	1641	12.6	201	18.0	80	8.4	4915	15.9	3093	9.8
2 of Middle Eas	484	3.1	854	4.6	370	3.1	406	3.1	130	11.6	66	7.0	2366	7.7	1286	4.1
Other M.East	2346	14.8	1843	9.8	920	7.6	1236	9.5	72	6.4	13	1.4	2549	8.3	1807	5.7
Oceania	9	0.1	27	0.1	-	-	0	0.0	11	1.0	7	0.8	694	2.3	895	2.8
Other Oceania	0	0.0	1	0.0	-	-	0	0.0	1	0.1	-	0.0	16	0.1	18	0.1
North America	1206	7.6	2755	14.7	66	0.5	662	5.1	•	•	•	•	4506	14.6	6297	19.9
N.America una.	0	0.0	1	0.0	-	-	-	-	•	•	•	•	-	0.0	0	0.0
Other America	141	0.9	229	1.2	132	1.1	126	1.0	352	31.5	345	36.4	884	2.9	1097	3.5
4 of O.America	52	0.3	61	0.3	1	0.0	0	0.0	200	17.9	240	25.3	347	1.1	471	1.5
Oth.Oth.America	89	0.6	168	0.9	131	1.1	126	1.0	153	13.6	104	11.0	537	1.7	626	2.0
Western Europe	7566	47.8	8150	43.5	3762	31.1	4281	32.8	140	12.6	205	21.7	607	2.0	582	1.8
EEC(9)	5704	36.0	6017	32.1	2065	17.1	2273	17.4	124	11.1	171	18.0	277	0.9	244	0.8
Northern Europe	989	6.2	943	5.0	334	2.8	274	2.1	4	0.4	7	0.7	100	0.3	87	0.3
Southern Europe	510	3.2	793	4.2	1056	8.7	1286	9.9	11	0.9	26	2.7	208	0.7	232	0.7
Central Europe	332	2.1	329	1.8	256	2.1	359	2.8	1	0.1	2	0.2	21	0.1	19	0.1
W.Europe unallo	32	0.2	68	0.4	51	0.4	89	0.7	0	0.0	-	0.0	-	0.0	0	0.0
Eastern Europe	1937	12.2	2199	11.7	1762	14.6	1619	12.4	2	0.2	1	0.1	2060	6.7	2098	6.6
7 of E.Europe	645	4.1	755	4.0	-	•	-	-	1	0.1	-	0.0	48	0.2	26	0.1
USSR	1291	8.2	1444	7.7	1762	14.6	1619	12.4	1	0.1	1	0.1	2012	6.5	2072	6.5
E.Europe unallo	-	-	-	-	-	-	-	-	-	-	-	-	-	-	-	-
Unallocated	2	0.0	0	0.0	4323	35.7	3735	28.6	-	-	-	-	-	-	-	-
World	15827	100.0	18741	100.0	12093	100.0	13045	100.0	1118	100.0	948	100.0	30821	100.0	31684	100.0

Table II.28. Changes in exports of all products by region or country

Importers	Oceania			
Years	1983		1984	
	1000t	%	1000t	%
Africa	2	0.2	1	0.1
North Africa	-	-	-	-
South Africa	0	0.0	0	0.0
Other Africa	2	0.2	0	0.0
Far East	539	54.8	570	60.9
3 of Far East	193	19.6	269	28.7
Japan	50	5.0	34	3.6
Other Far East	297	30.2	267	28.5
Middle East	94	9.5	108	11.6
2 of Middle East	43	4.4	1	0.1
Other M.East	51	5.1	107	11.5
Oceania	.	.	.	.
Other Oceania	.	.	.	.
North America	177	18.0	204	21.8
N.America una.	0	0.0	-	-
Other America	28	2.8	22	2.3
4 of O.America	21	2.1	21	2.3
Oth.Oth.America	7	0.7	1	0.1
Western Europe	115	11.7	1	0.1
EEC(9)	51	5.2	1	0.1
Northern Europe	-	-	0	0.0
Southern Europe	64	6.5	0	0.0
Central Europe	0	0.0	0	0.0
W.Europe unallo	-	-	-	-
Eastern Europe	-	-	-	-
7 of E.Europe	-	-	-	-
USSR	-	-	-	-
E.Europe unallo	-	-	-	-
Unallocated	28	2.9	30	3.2
World	984	100.0	936	100.0

Table II.29. Changes in exports of long products by region or country

Importers	Economic European community				North European countries				South European countries				Central European countries			
Years	1983		1984		1983		1984		1983		1984		1983		1984	
	1000t	%	1000t	%	1000t	%	1000t	%	1000t	%	1000t	%	1000t	%	1000t	%
Africa	832	3.1	772	2.5	25	0.7	22	0.5	1135	14.7	999	10.8	32	0.9	59	1.5
North Africa	486	1.8	485	1.5	20	0.6	18	0.5	998	12.9	633	6.8	31	0.9	58	1.5
South Africa	13	0.0	19	0.1	1	0.0	1	0.0	0	0.0	1	0.0	1	0.0	1	0.0
Other Africa	333	1.2	268	0.9	4	0.1	2	0.1	136	1.8	365	4.0	1	0.0	0	0.0
Far East	718	2.6	1082	3.5	27	0.8	50	1.2	67	0.9	432	4.7	8	0.2	11	0.3
3 of Far East	291	1.1	673	2.1	9	0.2	41	1.0	50	0.7	317	3.4	4	0.1	9	0.2
Japan	2	0.0	2	0.0	1	0.0	1	0.0	-	-	1	0.0	0	0.0	0	0.0
Other Far East	424	1.6	407	1.3	17	0.5	7	0.2	16	0.2	114	1.2	4	0.1	2	0.1
Middle East	828	3.0	662	2.1	32	0.9	22	0.5	1860	24.1	1782	19.3	13	0.4	41	1.0
2 of Middle East	282	1.0	189	0.6	15	0.4	10	0.2	356	4.6	696	7.5	1	0.0	4	0.1
Other M.East	547	2.0	473	1.5	17	0.5	12	0.3	1504	19.5	1087	11.8	13	0.4	38	1.0
Oceania	28	0.1	43	0.1	2	0.1	3	0.1	0	0.0	0	0.0	2	0.1	2	0.1
Other Oceania	9	0.0	16	0.1	0	0.0	-	-	-	-	0	0.0	-	-	-	-
North America	1043	3.8	1402	4.5	57	1.6	129	3.2	264	3.4	623	6.7	6	0.2	24	0.6
N.America una.	3	0.0	3	0.0	-	-	-	-	0	0.0	0	0.0	-	-	-	-
Other America	239	0.9	299	1.0	5	0.1	8	0.2	32	0.4	77	0.8	8	0.2	9	0.2
4 of O.America	73	0.3	159	0.5	4	0.1	5	0.1	4	0.1	5	0.1	6	0.2	1	0.0
Oth.Oth.America	166	0.6	139	0.4	1	0.0	3	0.1	29	0.4	72	0.8	2	0.1	8	0.2
Western Europe	1565	5.8	1703	5.4	770	21.7	852	20.9	840	10.9	878	9.5	685	20.0	837	21.2
EEC(9)	.	.	.	.	748	21.1	820	20.1	610	7.9	731	7.9	665	19.4	807	20.4
Northern Europe	446	1.6	534	1.7	15	0.4	17	0.4	147	1.9	77	0.8	6	0.2	7	0.2
Southern Europe	344	1.3	373	1.2	6	0.2	13	0.3	.	.	.	.	14	0.4	23	0.6
Central Europe	752	2.8	763	2.4	0	0.0	1	0.0	70	0.9	57	0.6	.	.	.	.
W.Europe unallo	24	0.1	34	0.1	0	0.0	1	0.0	13	0.2	13	0.1	0	0.0	0	0.0
Eastern Europe	502	1.8	366	1.2	74	2.1	55	1.4	540	7.0	647	7.0	105	3.1	60	1.5
7 of E.Europe	79	0.3	83	0.3	10	0.3	13	0.3	129	1.7	152	1.6	99	2.9	58	1.5
USSR	423	1.6	283	0.9	64	1.8	43	1.0	412	5.3	495	5.4	6	0.2	2	0.0
E.Europe unallo	-	-	-	-	-	-	-	-	-	-	-	-	-	-	-	-
Unallocated	2	0.0	2	0.0	-	-	-	-	0	0.0	0	0.0	-	-	-	-
World	5757	21.2	6331	20.2	991	27.9	1141	28.0	4739	61.5	5438	58.9	859	25.1	1045	26.4

Table II.30.   Changes in exports of long products by region or country

Importers	Oth West European countries 1983 1000t	1983 %	1984 1000t	1984 %	Eastern European countries 1983 1000t	1983 %	1984 1000t	1984 %	North American countries 1983 1000t	1983 %	1984 1000t	1984 %	Japan 1983 1000t	1983 %	1984 1000t	1984 %
Africa	1192	7.5	1080	5.8	44	0.4	65	0.5	32	2.8	17	1.8	46	0.1	30	0.1
North Africa	1049	6.6	709	3.8	21	0.2	21	0.2	22	1.9	0	0.0	14	0.0	3	0.0
South Africa	2	0.0	3	0.0	-	-	-	-	0	0.0	0	0.0	5	0.0	10	0.0
Other Africa	141	0.9	368	2.0	24	0.2	44	0.3	10	0.9	16	1.7	28	0.1	18	0.1
Far East	102	0.6	493	2.6	106	0.9	337	2.6	22	2.0	28	3.0	4613	15.0	4551	14.4
3 of Far East	63	0.4	368	2.0	61	0.5	156	1.2	6	0.6	2	0.2	2427	7.9	3014	9.5
Japan	1	0.0	2	0.0	-	-	-	0.0	3	0.3	1	0.1	-	-	-	-
Other Far East	38	0.2	123	0.7	46	0.4	181	1.4	13	1.2	25	2.6	2186	7.1	1537	4.8
Middle East	1905	12.0	1846	9.9	981	8.1	1307	10.0	72	6.4	12	1.3	2166	7.0	1274	4.0
2 of Middle Eas	371	2.3	710	3.8	302	2.5	287	2.2	38	3.4	9	1.0	1320	4.3	513	1.6
Other M.East	1534	9.7	1137	6.1	679	5.6	1020	7.8	34	3.1	3	0.3	846	2.7	760	2.4
Oceania	4	0.0	6	0.0	-	-	0	0.0	0	0.0	1	0.1	122	0.4	168	0.5
Other Oceania	0	0.0	0	0.0	-	-	0	0.0	0	0.0	0	0.0	9	0.0	9	0.0
North America	327	2.1	777	4.1	33	0.3	140	1.1	.	.	.	.	1069	3.5	1549	4.9
N.America una.	0	0.0	0	0.0	-	-	-	-	.	.	.	.	-	-	-	-
Other America	45	0.3	94	0.5	41	0.3	37	0.3	88	7.9	109	11.5	96	0.3	96	0.3
4 of O.America	14	0.1	11	0.1	1	0.0	0	0.0	50	4.5	92	9.7	36	0.1	59	0.2
Oth.Oth.America	31	0.2	83	0.4	40	0.3	37	0.3	38	3.4	17	1.8	60	0.2	37	0.1
Western Europe	2682	16.9	3095	16.5	1011	8.4	1105	8.5	29	2.6	37	3.9	66	0.2	70	0.2
EEC(9)	2023	12.8	2358	12.6	545	4.5	665	5.1	22	2.0	31	3.2	40	0.1	45	0.1
Northern Europe	414	2.6	334	1.8	155	1.3	120	0.9	2	0.2	4	0.4	4	0.0	3	0.0
Southern Europe	88	0.6	251	1.3	211	1.7	202	1.5	4	0.4	2	0.2	22	0.1	21	0.1
Central Europe	143	0.9	138	0.7	96	0.8	82	0.6	1	0.0	1	0.1	0	0.0	0	0.0
W.Europe unallo	13	0.1	15	0.1	4	0.0	37	0.3	-	-	-	-	0	0.0	-	-
Eastern Europe	719	4.5	761	4.1	1067	8.8	1006	7.7	1	0.1	0	0.0	253	0.8	135	0.4
7 of E.Europe	237	1.5	223	1.2	-	-	-	-	-	-	-	-	2	0.0	0	0.0
USSR	482	3.0	539	2.9	1067	8.8	1006	7.7	1	0.1	0	0.0	251	0.8	135	0.4
E.Europe unallo	-	-	-	-	-	-	-	-	-	-	-	-	-	-	-	-
Unallocated	0	0.0	0	0.0	1876	15.5	1453	11.1	-	-	-	-	-	-	-	-
World	6976	44.1	8152	43.5	5160	42.7	5451	41.8	244	21.9	205	21.6	8432	27.4	7873	24.8

Table II.31. Changes in exports of long products by region or country

Years	Oceania			
	1983		1984	
Importers	1000t	%	1000t	%
Africa	1	0.1	0	0.0
North Africa	–	–	–	–
South Africa	0	0.0	0	0.0
Other Africa	1	0.1	–	–
Far East	89	9.1	74	7.9
3 of Far East	56	5.7	54	5.8
Japan	–	–	0	0.0
Other Far East	34	3.4	20	2.1
Middle East	39	3.9	0	0.0
2 of Middle Eas	35	3.5	–	–
Other M.East	4	0.4	0	0.0
Oceania	.	.	.	.
Other Oceania	.	.	.	.
North America	23	2.3	11	1.1
N.America una.	–	–	–	–
Other America	7	0.7	0	0.0
4 of O.America	–	–	–	–
Oth.Oth.America	7	0.7	0	0.0
Western Europe	4	0.4	0	0.0
EEC(9)	4	0.4	0	0.0
Northern Europe	–	–	–	–
Southern Europe	–	–	–	–
Central Europe	–	–	0	0.0
W.Europe unallo	–	–	–	–
Eastern Europe	–	–	–	–
7 of E.Europe	–	–	–	–
USSR	–	–	–	–
E.Europe unallo	–	–	–	–
Unallocated	–	–	0	0.0
World	163	16.5	85	9.1

Table II.32. Changes in exports of flat products by region or country

Importers	Economic European community 1983 1000t	1983 %	1984 1000t	1984 %	North European countries 1983 1000t	1983 %	1984 1000t	1984 %	South European countries 1983 1000t	1983 %	1984 1000t	1984 %	Central European countries 1983 1000t	1983 %	1984 1000t	1984 %
Africa	606	2.2	643	2.1	4	0.1	6	0.1	50	0.6	69	0.7	2	0.0	2	0.0
North Africa	275	1.0	326	1.0	2	0.0	2	0.0	47	0.6	64	0.7	1	0.0	1	0.0
South Africa	22	0.1	35	0.1	1	0.0	2	0.0	0	0.0	0	0.0	0	0.0	0	0.0
Other Africa	308	1.1	281	0.9	2	0.0	2	0.0	3	0.0	4	0.0	0	0.0	1	0.0
Far East	1391	5.1	1338	4.3	34	1.0	45	1.1	186	2.4	252	2.7	1	0.0	1	0.0
3 of Far East	950	3.5	975	3.1	13	0.4	17	0.4	110	1.4	84	0.9	1	0.0	5	0.1
Japan	3	0.0	7	0.0	3	0.1	3	0.1	38	0.5	106	1.1	0	0.0	4	0.1
Other Far East	438	1.6	356	1.1	18	0.5	24	0.6	38	0.5	62	0.7	0	0.0	0	0.0
Middle East	814	3.0	854	2.7	39	1.1	7	0.2	198	2.6	209	2.3	47	1.4	43	1.1
2 of Middle Eas	170	0.6	185	0.6	5	0.1	5	0.1	36	0.5	54	0.6	10	0.3	14	0.4
Other M.East	644	2.4	669	2.1	35	1.0	2	0.0	161	2.1	155	1.7	37	1.1	28	0.7
Oceania	32	0.1	40	0.1	2	0.1	4	0.1	-	0.0	12	0.1	0	0.0	0	0.0
Other Oceania	8	0.0	8	0.0	-	-	0	0.0	-	-	1	0.0	-	-	-	-
North America	1516	5.6	1910	6.1	307	8.7	409	10.1	285	3.7	556	6.0	10	0.3	25	0.6
N.America una.	0	0.0	1	0.0	-	-	-	-	1	-	1	0.0	-	-	-	-
Other America	343	1.3	463	1.5	6	0.2	9	0.2	53	0.7	47	0.5	1	0.0	4	0.1
4 of O.America	143	0.5	216	0.7	5	0.1	7	0.2	20	0.3	22	0.2	1	0.0	3	0.1
Oth.Oth.America	200	0.7	247	0.8	1	0.0	2	0.0	33	0.4	25	0.3	0	0.0	1	0.0
Western Europe	3202	11.8	3492	11.2	1142	32.2	1089	26.7	350	4.5	365	4.0	993	29.0	954	24.1
EEC(9)	-	-	-	-	1044	29.4	1032	25.4	302	3.9	309	3.3	869	25.4	826	20.9
Northern Europe	1100	4.0	1188	3.8	54	1.5	29	0.7	19	0.2	19	0.2	48	1.4	52	1.3
Southern Europe	1060	3.9	1222	3.9	43	1.2	27	0.7	-	-	-	-	76	2.2	76	1.9
Central Europe	1024	3.8	1058	3.4	0	0.0	1	0.0	28	0.4	35	0.4	-	-	-	-
W.Europe unallo	18	0.1	24	0.1	61	1.7	62	1.5	0	0.0	2	0.0	-	-	-	-
Eastern Europe	1912	7.0	2288	7.3	23	0.6	24	0.6	101	1.3	71	0.8	584	17.1	668	16.9
7 of E.Europe	214	0.8	228	0.7	38	1.1	38	0.9	62	0.8	59	0.6	161	4.7	157	4.0
USSR	1697	6.2	2061	6.6	-	-	-	-	38	0.5	12	0.1	424	12.4	511	12.9
E.Europe unallo	-	-	-	-	-	-	-	-	-	-	-	-	-	-	-	-
Unallocated	8	0.0	6	0.0	-	-	-	-	-	-	0	0.0	-	-	-	-
World	9823	36.2	11035	35.3	1596	45.0	1630	40.0	1222	15.9	1581	17.1	1639	47.8	1700	43.0

Table II.33.   Changes in exports of flat products by region or country

Importers	Oth West European countries 1983 1000t	%	1984 1000t	%	Eastern European countries 1983 1000t	%	1984 1000t	%	North American countries 1983 1000t	%	1984 1000t	%	Japan 1983 1000t	%	1984 1000t	%
Africa	56	0.4	76	0.4	7	0.1	8	0.1	10	0.9	4	0.4	359	1.2	328	1.0
North Africa	50	0.3	67	0.4	0	0.0	2	0.0	4	0.3	0	0.0	13	0.0	7	0.0
South Africa	1	0.0	2	0.0	-	-	-	-	1	0.1	0	0.0	43	0.1	54	0.2
Other Africa	5	0.0	7	0.0	7	0.1	6	0.1	6	0.5	3	0.3	304	1.0	267	0.8
Far East	222	1.4	302	1.6	491	4.1	407	3.1	286	25.6	226	23.8	7922	25.7	7995	25.2
3 of Far East	124	0.8	105	0.6	210	1.7	195	1.5	60	5.3	52	5.5	4509	14.6	4827	15.2
Japan	41	0.3	110	0.6	109	0.9	120	0.9	4	0.3	5	0.5	-	-	-	-
Other Far East	56	0.4	87	0.5	172	1.4	91	0.7	222	19.9	168	17.8	3414	11.1	3168	10.0
Middle East	284	1.8	258	1.4	195	1.6	240	1.8	39	3.5	24	2.5	1466	4.8	940	3.0
2 of Middle Eas	51	0.3	73	0.4	57	0.5	106	0.8	26	2.3	19	2.1	364	1.2	316	1.0
Other M.East	233	1.5	185	1.0	138	1.1	134	1.0	13	1.2	4	0.4	1101	3.6	624	2.0
Oceania	2	0.0	16	0.1	-	-	-	-	4	0.4	2	0.2	370	1.2	524	1.7
Other Oceania	-	-	1	0.0	-	-	-	-	0	0.0	0	0.0	6	0.0	8	0.0
North America	602	3.8	990	5.3	22	0.2	472	3.6	·	-	·	·	2212	7.2	2938	9.3
N.America una.	-	-	1	0.0	-	-	-	-	137	12.2	138	14.6	0	0.0	-	-
Other America	60	0.4	59	0.3	72	0.6	60	0.5	89	8.0	97	10.2	469	1.5	615	1.9
4 of O.America	26	0.2	32	0.2	1	0.0	0	0.0	47	4.2	41	4.3	124	0.4	196	0.6
Oth.Oth.America	35	0.2	28	0.1	72	0.6	60	0.5	79	7.1	121	12.8	344	1.1	419	1.3
Western Europe	2918	18.4	2864	15.3	1652	13.7	1796	13.8	75	6.7	98	10.3	263	0.9	210	0.7
EEC(9)	2216	14.0	2168	11.6	928	7.7	1035	7.9	1	0.1	1	0.1	128	0.4	99	0.3
Northern Europe	367	2.3	386	2.1	142	1.2	127	1.0	3	0.3	22	2.3	24	0.1	16	0.1
Southern Europe	196	1.2	175	0.9	437	3.6	413	3.2	0	0.0	1	0.1	92	0.3	80	0.3
Central Europe	139	0.9	132	0.7	123	1.0	194	1.5	-	-	-	-	19	0.1	15	0.0
W.Europe unallo	1	0.0	3	0.0	21	0.2	28	0.2	0	0.0	0	0.0	1	0.0	0	0.0
Eastern Europe	746	4.7	801	4.3	235	1.9	127	1.0	0	0.0	0	0.0	512	1.7	493	1.6
7 of E.Europe	246	1.6	240	1.3	-	-	-	-	0	0.0	0	0.0	13	0.0	9	0.0
USSR	500	3.2	562	3.0	235	1.9	127	1.0	-	-	-	-	500	1.6	484	1.5
E.Europe unallo	-	-	-	-	-	-	-	-	-	-	-	-	-	-	-	-
Unallocated	-	-	0	0.0	1634	13.5	1751	13.4	-	-	-	-	-	-	-	-
World	4890	30.9	5367	28.6	4309	35.6	4861	37.3	555	49.6	514	54.3	13574	44.0	14044	44.3

Table II.34. Changes in exports of flat products by region or country

Years	Oceania			
	1983		1984	
Importers	1000t	%	1000t	%
Africa	1	0.1	0	0.0
North Africa	-	-	-	-
South Africa	-	-	0	0.0
Other Africa	1	0.1	0	0.0
Far East	238	24.2	172	18.3
3 of Far East	117	11.9	48	5.1
Japan	0	0.0	0	0.0
Other Far East	121	12.3	124	13.2
Middle East	10	1.0	2	0.2
2 of Middle Eas	3	0.3	1	0.1
Other M.East	7	0.7	1	0.1
Oceania	.	.	.	.
Other Oceania	.	.	.	.
North America	141	14.4	185	19.8
N.America una.	-	-	-	-
Other America	1	0.1	1	0.1
4 of O.America	0	0.0	0	0.0
Oth.Oth.America	0	0.0	1	0.1
Western Europe	51	5.2	0	0.0
EEC(9)	38	3.9	0	0.0
Northern Europe	-	-	-	-
Southern Europe	13	1.3	0	0.0
Central Europe	-	-	-	-
W.Europe unallo	-	-	-	-
Eastern Europe	-	-	-	-
7 of E.Europe	-	-	-	-
USSR	-	-	-	-
E.Europe unallo	-	-	-	-
Unallocated	-	-	-	-
World	442	45.0	359	38.4

Table II.35. Changes in exports of tubes and fittings by region or country

Importers / Years	Economic European community 1983 1000t	1983 %	1984 1000t	1984 %	North European countries 1983 1000t	1983 %	1984 1000t	1984 %	South European countries 1983 1000t	1983 %	1984 1000t	1984 %	Central European countries 1983 1000t	1983 %	1984 1000t	1984 %
Africa	320	1.2	365	1.2	5	0.1	8	0.2	92	1.2	51	0.5	11	0.3	10	0.3
North Africa	171	0.6	207	0.7	1	0.0	2	0.0	78	1.0	32	0.3	10	0.3	9	0.2
South Africa	29	0.1	23	0.1	2	0.1	5	0.1	1	0.0	2	0.0	0	0.0	0	0.0
Other Africa	121	0.4	135	0.4	2	0.1	2	0.0	13	0.2	16	0.2	1	0.0	1	0.0
Far East	547	2.0	492	1.6	9	0.2	9	0.2	65	0.8	53	0.6	1	0.0	3	0.1
3 of Far East	450	1.7	430	1.4	4	0.1	5	0.1	52	0.7	51	0.6	0	0.0	2	0.1
Japan	1	0.0	1	0.0	1	0.0	1	0.0	-	-	-	-	0	0.0	0	0.0
Other Far East	97	0.4	61	0.2	3	0.1	3	0.1	13	0.2	2	0.0	0	0.0	1	0.0
Middle East	626	2.3	615	2.0	4	0.1	3	0.1	227	3.0	256	2.8	7	0.2	13	0.3
2 of Middle East	309	1.1	290	0.9	1	0.0	1	0.0	14	0.2	13	0.1	1	0.0	3	0.1
Other M.East	317	1.2	325	1.0	3	0.1	2	0.1	213	2.8	243	2.6	6	0.2	10	0.3
Oceania	59	0.2	15	0.0	2	0.0	3	0.1	1	0.0	1	0.0	0	0.0	0	0.0
Other Oceania	4	0.0	6	0.0	0	0.0	-	-	0	0.0	0	0.0	0	0.0	0	0.0
North America	540	2.0	1368	4.4	22	0.6	46	1.1	79	1.0	321	3.5	4	0.1	82	2.1
N.America una.	1	0.0	1	0.0	-	-	-	-	0	0.0	0	0.0	-	-	-	-
Other America	147	0.5	107	0.3	5	0.2	5	0.1	12	0.2	12	0.1	1	0.0	0	0.0
4 of O.America	118	0.4	80	0.3	3	0.1	4	0.1	6	0.1	2	0.0	0	0.0	0	0.0
Oth.Oth.America	29	0.1	27	0.1	2	0.1	1	0.0	6	0.1	10	0.1	0	0.0	0	0.0
Western Europe	978	3.6	905	2.9	190	5.4	204	5.0	140	1.8	193	2.1	285	8.3	331	8.4
EEC(9)	644	2.4	534	1.7	175	4.9	190	4.7	116	1.5	161	1.7	246	7.2	290	7.3
Northern Europe	116	0.4	104	0.3	.	0.0	.	0.0	17	0.2	21	0.2	34	1.0	36	0.9
Southern Europe	207	0.8	259	0.8	5	0.1	3	0.1	.	.	.	0.1	5	0.1	5	0.1
Central Europe	11	0.0	8	0.0	10	0.3	11	0.3	7	0.1	10	0.1	.	.	.	.
W.Europe unallo	-	-	-	-	0	0.0	0	0.0	0	0.0	2	0.0	.	.	.	.
Eastern Europe	2870	10.6	2757	8.8	14	0.4	12	0.3	157	2.0	188	2.0	196	5.7	209	5.3
7 of E.Europe	191	0.7	237	0.8	5	0.1	3	0.1	54	0.7	52	0.6	11	0.3	13	0.3
USSR	2679	9.9	2520	8.1	9	0.3	9	0.2	102	1.3	135	1.5	184	5.4	197	5.0
E.Europe unallo	-	-	-	-	-	-	-	-	-	-	-	-	-	-	-	-
Unallocated	2	0.0	3	0.0	-	-	-	-	-	-	0	-	-	-	-	-
World	6091	22.4	6629	21.2	251	7.1	292	7.2	773	10.0	1074	11.6	505	14.7	649	16.4

Table II.36. Changes in exports of tubes and fittings by region or country

Importers	Oth West European countries 1983 1000t	%	1984 1000t	%	Eastern European countries 1983 1000t	%	1984 1000t	%	North American countries 1983 1000t	%	1984 1000t	%	Japan 1983 1000t	%	1984 1000t	%
Africa	108	0.7	69	0.4	6	0.0	8	0.1	10	0.9	18	1.9	117	0.4	103	0.3
North Africa	88	0.6	42	0.2	4	0.0	0	0.0	2	0.2	5	0.5	40	0.1	27	0.1
South Africa	4	0.0	7	0.0	-	-	-	-	1	0.1	1	0.1	39	0.1	36	0.1
Other Africa	16	0.1	19	0.1	1	0.0	8	0.1	7	0.7	13	1.4	39	0.1	40	0.1
Far East	75	0.5	65	0.3	20	0.2	49	0.4	31	2.8	11	1.1	1801	5.8	2160	6.8
3 of Far East	57	0.4	58	0.3	18	0.1	38	0.3	20	1.8	4	0.5	1094	3.5	1481	4.7
Japan	1	0.0	1	0.0	-	-	-	-	1	0.1	1	0.1	-	-	-	-
Other Far East	17	0.1	6	0.0	2	0.0	10	0.1	10	0.9	5	0.6	707	2.3	679	2.1
Middle East	238	1.5	272	1.5	37	0.3	20	0.2	83	7.4	43	4.5	786	2.5	561	1.8
2 of Middle Eas	17	0.1	17	0.1	-	-	1	0.0	65	5.8	37	3.9	479	1.6	272	0.9
Other M.East	222	1.4	255	1.4	37	0.3	19	0.1	19	1.7	6	0.6	307	1.0	289	0.9
Oceania	2	0.0	4	0.0	-	-	0	0.0	6	0.6	4	0.4	117	0.4	86	0.3
Other Oceania	0	0.0	0	0.0	-	-	-	-	1	0.1	0	0.0	1	0.0	1	0.0
North America	104	0.7	449	2.4	9	0.1	44	0.3	.	.	.	.	805	2.6	1325	4.2
N.America una.	0	0.0	0	0.0	-	-	-	-	.	.	.	.	-	-	0	0.0
Other America	19	0.1	17	0.1	15	0.1	24	0.2	63	5.6	65	6.8	141	0.5	207	0.7
4 of O.America	10	0.1	6	0.0	0	0.0	-	-	35	3.2	36	3.8	36	0.1	67	0.2
Oth.Oth.America	9	0.1	11	0.1	15	0.1	24	0.2	28	2.5	29	3.0	104	0.3	140	0.4
Western Europe	744	4.7	883	4.7	170	1.4	211	1.6	13	1.2	16	1.7	150	0.5	173	0.5
EEC(9)	537	3.4	641	3.4	88	0.7	124	0.9	10	0.9	12	1.2	70	0.2	97	0.3
Northern Europe	157	1.0	168	0.9	36	0.3	27	0.2	1	0.1	2	0.2	62	0.2	57	0.2
Southern Europe	14	0.1	32	0.2	22	0.2	26	0.2	2	0.2	2	0.2	17	0.1	18	0.1
Central Europe	35	0.2	40	0.2	24	0.2	33	0.3	0	0.0	0	0.0	1	0.0	1	0.0
W.Europe unallo	1	0.0	2	0.0	0	0.0	1	0.0	0	0.0	0	0.0	0	0.0	1	0.1
Eastern Europe	367	2.3	409	2.2	429	3.5	447	3.4	1	0.1	1	0.1	1283	4.2	1468	4.6
7 of E.Europe	71	0.4	68	0.4	-	-	-	-	1	0.1	0	0.0	33	0.1	17	0.1
USSR	296	1.9	341	1.8	429	3.5	447	3.4	0	0.0	1	0.1	1250	4.1	1451	4.6
E.Europe unallo	-	-	-	-	-	-	-	-	-	-	-	-	-	-	-	-
Unallocated	-	-	0	0.0	345	2.9	371	2.8	-	-	-	-	-	-	-	-
World	1658	10.5	2169	11.6	1032	8.5	1174	9.0	208	18.6	158	16.6	5200	16.9	6083	19.2

Table II.37. Changes in exports of tubes and fittings by region or country

Years	Oceania			
	1983		1984	
Importers	1000t	%	1000t	%
Africa	0	0.0	0	0.0
North Africa	-	-	-	-
South Africa	0	0.0	0	0.0
Other Africa	-	-	0	0.0
Far East	1	0.1	1	0.2
3 of Far East	0	0.0	0	0.0
Japan	0	0.0	-	-
Other Far East	1	0.1	1	0.1
Middle East	0	0.0	0	0.0
2 of Middle Eas	0	0.0	0	0.0
Other M.East	0	0.0	0	0.0
Oceania	.	.	.	.
Other Oceania	.	.	.	.
North America	0	0.0	0	0.1
N.America una.	0	0.0	-	-
Other America	0	0.0	-	-
4 of O.America	-	-	-	-
Oth.Oth.America	0	0.0	-	-
Western Europe	0	0.0	-	-
EEC(9)	0	0.0	-	-
Northern Europe	-	-	-	-
Southern Europe	-	-	-	-
Central Europe	-	-	-	-
W.Europe unallo	-	-	-	-
Eastern Europe	-	-	-	-
7 of E.Europe	-	-	-	-
USSR	-	-	-	-
E.Europe unallo	-	-	-	-
Unallocated	28	2.9	30	3.2
World	29	3.0	32	3.4

Printed at United Nations, Geneva
GE.87-22700/0233G/0264G
October 1987—2,570

02300P

United Nations publication
Sales No. E.87.II.E.33

ISBN 92-1-116408-7